THE WOMAN HOBBY FARMER

Karen Lanier

The Woman Hobby Farmer

CompanionHouse Books™ is an imprint of Fox Chapel Publishing, Inc.

Project Team
Director of Product Development and Editorial Operations: Christopher Reggio
Editor: Amy Deputato
Copy Editor: Laura Taylor
Design: Mary Ann Kahn
Index: Elizabeth Walker

Library of Congress Cataloging-in-Publication Data
Names: Lanier, Karen, 1975- author.
Title: The woman hobby farmer : female guidance for growing food, raising livestock, and building a farm-based business / Karen Lanier.
Description: East Petersburg, PA : CompanionHouse Books, [2017] | Includes bibliographical references and index.
Identifiers: LCCN 2017013667 (print) | LCCN 2017020513 (ebook) | ISBN 9781620082614 (ebook) | ISBN 9781620082607 (softcover)
Subjects: LCSH: Women farmers.
Classification: LCC HD6077 (ebook) | LCC HD6077 .L36 2017 (print) | DDC 630.68--dc23
LC record available at https://lccn.loc.gov/2017013667

Fox Chapel Publishing
903 Square Street
Mount Joy, PA 17552

Fox Chapel Publishers International Ltd.
7 Danefield Road, Selsey (Chichester)
West Sussex PO20 9DA, U.K.

www.facebook.com/companionhousebooks

Printed and bound in China
20 19 18 17   2 4 6 8 10 9 7 5 3 1

CONTENTS

Introduction 4

Chapter 1: Assessments 9

Chapter 2: Taking Care of Your Farm 69

Chapter 3: Taking Care of Yourself.... 121

Chapter 4: Harvest and Share the Bounty.... 153

Chapter 5: Good Teacher, Good Student.... 173

Chapter 6: Lessons Learned 193

Chapter 7: Integrating Farm and Life 211

References and Recommended Reading.... 230

Index 232

Photo Credits 239

About the Author 240

INTRODUCTION

WELCOME, WOMEN

This book is a place to explore the feminine side of working with the land. It is for any gender and any level of farmer or gardener. The gentle wisdom shared here penetrates across boundaries. Why, then, do we need a book about women farmers?

When I first began researching for this book, I delved into the gender issue because I didn't want to make assumptions about who is biased against whom, what feminism is all about, and why being female makes someone a minority in the field of agriculture. If you take a quick look at the list of references at the back of this book, you'll see that my research led me to exploring gender identity. With farming, an endeavor that can be deeply soul-fulfilling at one moment and can bring up shame and guilt the next, it helps if we gain a deeper understanding of women in agriculture and build supportive relationships to do this work well. Farming requires your whole self, and the more you can see that self for what it is rather than what others have labeled it as, the truer you can be to your calling. It doesn't take much for food to grow, but consciously planting the social and environmental seeds for future generations of female farmers—and for all human beings—to flourish will be truly, deeply, and genuinely nourishing.

In my quest to understand and connect with more women who work the land, I interviewed farmers and gardeners: some I've known for years, some I've just recently met, some I only spoke to long-distance, and one I've known my entire life. I also attended the Southeast Wise Women's Herbal Conference in Black Mountain, North Carolina. Immersed in the diversity I found there, I forgot that I was around only women. I heard their stories and learned from their generations of accumulated knowledge. In a place where trust is assumed, it is so much easier to share and learn. This is the power of finding your community and settling into it. This book will give you some ideas about how to create or join a trusting community where you can give and receive food, medicine, and knowledge.

Living in a bubble of love and protection where wise women nurture and teach us is possible at times, yet we also have to deal with the uncomfortable realities of going it alone, getting confused with conflicting information, and facing fears that threaten to stop us from following our desire to grow food and/or raise animals. Progressing in any chosen path calls for crossing boundaries, dealing with discomfort, and stepping up into a leadership role when it's called for. In this book, we'll take a look at how some of our farming sisters handle challenges and gain confidence from understanding that many of these barriers are invisible, imagined, and easily dissolved.

If you look around at those who inspire you, what are they doing with their lives? When you think about your future, does it look like a typical retirement? When you assess the talents and skills you have accumulated thus far, can you imagine unique combinations that don't necessarily fit into a box? We all have so much to offer and the potential to grow in healthy ways. Identifying your personal boundaries, where you won't compromise, can actually open up an immense freedom to explore the ways you can live every day to its fullest. At the risk of sounding like a self-help author, I want to encourage you to always keep your values and intentions a strong priority, no matter what work you do with your life. Our work does not define who we are. We define the meaning of our work.

There is no "normal" farm anymore. Regardless of widespread industrial factory farming, a new wave is rising throughout cities, suburbs, and the countryside. Think permaculture, agroforestry, aquaponics, urban vertical farming. A farm is anything but a series of straight lines, and the same goes for a farmer (curves are nice, aren't they?). The

farm of the 1950s did not look like the farm of the 1850s, and we can define what the farm of 2050 becomes. We are poised at an exciting time in which anything is possible. Hemp or hazelnuts can power our machines. Farmers can sell oak logs inoculated with shiitake spores, and customers can be co-owners and harvest their own food. Scarcity does not define our market or our future. Nothing is set in stone, and everything can be as fluid as you choose or as nature intends. Listen to what the land wants. She would like this to be a continuing conversation with you.

This book does not attempt to provide the specific instructions on how to plant, raise, harvest, preserve, and sell. Farming, whether for fun or profit, and its diverse interpretations are much too expansive to fit in any single book. Many great resources exist to provide specific instructions on growing crops and raising livestock. The evergreen advice of "consult your local agricultural extension agent" is one very valuable tip that I can pass on to you. Not only will the knowledge of your local cooperative extension cost you less than this book, it will be exactly what you need for your growing zone and microclimate. There are volunteer Master Gardeners in every state just waiting to lend you a hand, whether you are in a rural or urban setting.

In this book, you will find a wealth of wisdom, the kind that normally doesn't come in written form but through years of mentorship, trial and error, and hands-on learning. Without those types of knowledge, growth in your garden or in your soul will not amount to much.

If you are a female who is interested in supporting a feminine approach to agriculture, curious about what women's strengths as farmers are, or drawn to the idea of balancing all aspects of your inner and outer self, this book has something for you. You need not be a hobby farmer, a serious gardener, or a woman to effect change in your own body, community, and environment. The women in this book bring you opportunities to explore the hidden mysteries of natural cycles and our intuitive connections that help us get in sync with nature and with each other to grow food and to grow personally with ease.

Chapter 1

Assessments

"Idealism is good and you want to hold onto that to a degree, because that's how you make change in the world and achieve your dreams. But you also have to have some structure to it."

—Jessica Ballard, GreenHouse17

FIRST STEPS: GETTING REAL

Planning is an action step. Never underestimate the importance of thinking through a task, a project, an interest, even a small whim. Daydreams belong in the toolbox, along with phone numbers, recipes, spades, and hoes. Farming on any scale is about bringing a seed to fruition, be it a literal seed of a plant, or a metaphorical seed of an idea. Seeds of all kinds require time and nurturing. They cannot be hurried along until they are ready and the conditions are right, and they also can be surprisingly forgiving, tough, and resilient when the odds are stacked against them.

No matter where you are along the spectrum of farming experience, all of the stages this chapter will explore are worth looking into. Rather than thinking of any type of life experience as a linear progression from point A to point B, try to view these phases as spiraling around, and you can find yourself recognizing your plans and projects as fitting various points along these cycles. You can (and really ought to) revisit the earlier stages occasionally to see how you can refresh your point of view, especially in times when you may feel stuck or particularly challenged.

The exercises in this section begin with a quick, shoot-from-the-hip question to prompt your intuition. Jot down your ideas in the space provided on the pages, or designate a specific farming journal. Thinking and analyzing will come in due time. Along the way, absorb the examples and stories to stimulate your imagination, start conversations, compare notes, or even see how not to do something.

There are as many different ways to farm as there are farmers. Everyone is doing it right to some extent. Some are doing it right for the greater good of a sustainable planet; others are working for the good of their own children's future. Some may be doing it to benefit their retirement fund, while their neighbors may be doing it to honor the legacy of the generations before them. These reflect the personal and relationship values that guide decisions on a daily basis. The more conscious and deliberate we are about directing our actions to be aligned with our values, the more likely we are to reach our goals and live a fulfilling life. This applies to anyone, farmer or not.

PLANNING: WANT, KNOW, AND HAVE

To put it into terms that relate directly to your farm, this chapter takes you through a simple process of assessing your situation and identifying what is important to you. In short, we'll discover the answers to three important questions:

- What do you want?
- What do you know?
- What do you have?

The overall goal is to want what you have and to have what you want. Our farmer stories and your self-evaluations will help you discover ways to move in that direction.

WANT

What do you want?

This question is intentionally vague. Think and feel broadly, and don't limit yourself to ideas about farming or gardening. What do you really want in your life? Your heart's desires. Your gut instinct. Your longings and needs. If you are struggling to come up with anything, sit in a quiet place and listen to what bubbles to the surface. Take as long as you need.

What was the first thing you came up with? Write it down now. Don't think, just write. Complete sentences are not necessary. Doodles and drawings are welcome.

This is not a time to overthink and analyze. Your subconscious knows, so let it be free here. Daydreams and fantasies are allowed, but try to focus on your deepest and strongest impulses rather than passing whims. What has been nagging at you? What have you been ignoring? What do you know you would be capable of if only you had a chance to try it? If you need to let out a lot of ideas and emotions, scribble away. Again, take as long as you need.

Now close this book and let it sit for a day or two, but not longer than a week.

Welcome back. Take another look at your wants.

What pops out at you, rises to the surface, grabs your attention? Circle or highlight it. What did you write that you don't really want? Cross it out. Can any of them be combined? Are any contradicting each other? Are any a prerequisite for meeting another want, need, or desire? Pull out your main ideas and write them here:

REASSESS

Since we are cycle-based beings, nothing in our lives will remain static. Revisit this exercise and your notes annually, and give yourself the opportunity to reassess and shift your focus.

Look at these beautiful desires. They are what you value: your priorities, your hopes, and what you esteem as important and necessary. Carry them with you, leave notes to yourself, and send yourself text messages with these words. Make cute signs and decorate your walls with them. Just do whatever it takes to keep these precious gems in sight so that they guide you and remind you of what you truly want.

FARMING FOR ONE

If you are now trying to farm on your own, or if you just feel that way sometimes, finding a sense of confidence and belonging in a field that has traditionally revolved

CONNECT AND FIND SUPPORT

www.thegreenhorns.net

The Greenhorns are a fun and diverse group based in New York. They are reaching the nontraditional farming crowd with their media and outreach events, and their collaborative and inclusive approach is catching on nationwide.

www.farmhack.org

Farm Hack is exactly what folks who need tools and machinery need to know about. It is an open-source community that primarily lives online, but its events are popping up in rural and urban farm sites everywhere.

www.wfan.org

Women Food and Ag Network (WFAN) provides a listing of regional networks to connect women in sustainable agriculture. (Psst! Hey, farmer men! If you want to avoid dating websites and aren't afraid of strong women, you may want to get involved in some of these networks as well.)

www.nal.usda.gov/afsic

The US Department of Agriculture's (USDA) Alternative Farming Systems Information Center (AFSIC) provides many resources for beginning farmers, including the hows and whys of running a small business. AFSIC also specifically addresses women and minorities in agriculture.

www.fsa.usda.gov

The USDA's Farm Service Agency (FSA) offers its Microloan Program, and its website states that it is designed to "serve the unique financial operating needs of beginning, niche, and the smallest of family farm operations."

https://nifa.usda.gov/Extension

Your local agricultural cooperative extension office keeps tabs on the grants available to small farmers. Watch for educational opportunities designed not only to teach you the latest on topics such as ecological pest management but also to help you meet supportive farmers in your community.

around patriarchal families can be particularly challenging. The good news is that you are not alone.

Jessica Ballard made a decision to begin college as a single mother who found her passion in farming. Her favorite college professor assigned Stephen Covey's *The 7 Habits of Highly Effective People*. For extra credit, Jessica was tasked with writing a mission statement for herself with goals for the next five to ten years. Jessica remembers, "All the other students turned in theirs in the next class. I took the whole semester. It was so important to me to be intentional."

Jessica's passion for farming led her to help other women through working the land.

She sat by herself and pondered why she wanted to farm. Her journey had begun in culinary school, and she felt deeply drawn to caring for the earth as well as wanting people to eat well. She says now, "I wrote out little goals, like get off food stamps, find a community garden to work in, maintain an agricultural job. Everything I wrote ten years ago has come about." She is glad she had a professor who encouraged her to create a strategy to move toward her goals.

Storey Slone readily admits, "I was a really rebellious young person. My foray into agriculture began with figuring out how to not be dependent on others, the government, and consumerism." Like many new farmers, Storey is a single female who is not inheriting any family farm. She has tried hard to bravely go it alone, and she recently made the decision to finish her college life before starting her farm life. At age twenty-five, she has earned an associate's degree in sustainable agriculture, worked for three different single female farmers, started a gardening club with more than 100 members, and, for an entire year, actively looked for someone to farm with by posting ads and networking like crazy. Her breaking point may have been when she went to lease land from a man, and he said, "Is it just you? We're really looking for a couple."

Storey's decision was partially a financial one. "I honestly don't mind if I have to work a job if I get to come home to a farm, whether I'm single or with someone." At Sterling College in Vermont, her self-designed major in sustainable agriculture and integrated forestry is grooming her for a foundation in an outdoor career so she can support her farming habit. She plans to return to her home state of Kentucky when she finishes. "My

goal is not to get rich farming. I really want to make Appalachian products and create a sustainable model for others."

Marlena Bolin credits her decision to become a farmer to a doctor who advised her to change her eating habits for the sake of her own health. She shifted away from processed foods and toward nutrient-dense vegetables. Her immune system improved tremendously. She reflects, "Perhaps it's a sense of duty that pushed me to pursue a career in farming. I wanted to contribute to the health of nearby groundwater, wildlife, and air quality."

Marlena rents two acres and grows more than thirty types of vegetables, specializing in heirloom tomatoes. Over her first five years in farming, she has discovered her deep roots in agriculture. Marlena says now, at age twenty-eight, that she is uncovering her heritage. "It wasn't until I took up farming did I learn of the rich family farm history that ran in my blood. I've been delighted to carry it on despite the lack of mentorship or the promise of inherited land or equipment."

She recounts a story that gets to the heart of what lone farmers fear most: sustaining an injury when nobody is around. "I got pinned between a tractor tire and a tiller once," she remembers. "It wasn't engaged, thankfully. I was able to shift the tiller away from me and squeeze out. I had a bruise from hell and didn't work at all the next day. I still have a tiny scar from the impact."

Marlena finds support in the Community Farm Alliance, Kentucky's grassroots network that does a great job of making friends for farmers. Workshops, conferences, and social events provide venues for formal partnerships and informal tailgate conversations. Learning about grants and programs offered by nearby universities has opened more doors for Marlena. She explains, "I'm opening up more land for additional CSA (community-supported agriculture) members and wholesale sales. I'm planning on hiring one full-time person to carry out major farm activities. I'm increasing my value-added products to [include] marinara, sun-dried tomatoes, and other shelf-stable goodies." Marlena is also planning on buying property in a few years where she can build some infrastructure, such as greenhouses, which renting won't allow.

Infrastructure is a common challenge that single farmers, new or experienced, deal with. In her mid-fifties, Susana Lein has been building up Salamander Springs Farm one block at a time, and she molded those blocks with her own two hands. Thirteen years after finding very affordable yet very inaccessible land, Susana has created a model off-the-grid sustainable lifestyle without going into debt.

Susana found a niche in growing heirloom popcorn, beans, herbs, and other vegetables. She says, "The corn has been one of my best and most famous things, but I didn't want to assume it ahead of starting." She started small and tried out a variety of crops. Now she teaches others to do the same.

With more than thirty years of farming experience, Susana is transitioning into teaching and writing by hosting permaculture and natural-building workshops as well as traveling to conferences. She says that the national network of organic, biodynamic, and permaculture organizations provides peers with whom she connects to exchange information and to grow personally.

Apprentices and WWOOFers (members of Worldwide Opportunities on Organic Farms) make up her seasonal family and labor force, and her reputation as a powerhouse is well known in the community near Berea, Kentucky. Susana credits this very community with keeping her sane through the trials and tribulations of farming solo. "People up and down the creek barter for things like raw milk and provide emotional support, which is so important. We're not an island, and you really can't farm solo." She continues, "Some people build up the land and don't build up the community. The only way to live sustainably is to share everybody's crops and everybody's talents."

Susana relies on her two capable hands and a supportive community.

RELATIONSHIPS AND FARMING

Charting your course is not so simple for women who aren't single. Identifying personal values and goals is only the first step. In a partnership, be it household or business, our individual needs and values must be balanced or supported by the relationship, an entity that exists with its own distinct principles and demands.

Rachael and Brent Dupree spontaneously bought a 50-acre farm just a few months before their wedding date. While they have their whole lives ahead of them to figure out what they want to do together on that farm, their preliminary discussions revealed separate but complementary values. Rachael explained, "I care more about cooking

Rachael and Brent on their wedding day.

and growing my own food, and he sees this property as a [financial] investment monetarily. We both see it as a happiness investment. I see it more as a way of life investment—how we can better the world—and I'm more dreamy about what it is we're doing."

I asked her if she would have done it alone. "Not 50 acres! I never expected it to be more than 5 or 10 acres. I would have gardened on my own, but it's better with another person. If he wasn't super excited about it, then it wouldn't happen."

Melissa Calhoun met her match when she was a young apprentice on an organic farm in Maine. The man who would become her husband was studying to be a farmer, and she fell for him while she was falling in love with the freshness of the garden. (This was Melissa's first time really tasting farm-fresh beets and peas.) Ten years later, growing challenges in the couple's divergent ideas on how to support a farm business culminated in their separation.

Many women hold down full-time jobs to support their partner's farm business, and in Melissa's case, she moved to a different state with her husband, who tried to make a serious business from organic farming. The couple bought property and, with three friends, established an intentional community. Melissa felt the strain of meeting her own needs in the midst of all this. "I felt like I was doing a lot of life changing, so income wasn't really on the top of my list. It never occurred to me that maybe I should make money to support his farming. Maybe it occurred to him…that [it] would have been nice for him to have a partner who supports him. I know a lot of farmers who do."

Melissa's story continues, now solo, three years after the divorce and the exodus of the members of her small community. She has the forested property to herself, where she practices herbalism with her own unique approach. Melissa muses, "People used to cut things down. Now it's only me, and I get to decide what lives and what gets moved away."

My Aunt Judy shared a great example of why it's important to discuss the values of both members in a partnership, not only in terms of an overall lifestyle but also in the way that they handle projects. They needed a fence. She and her six children were at

home on the farm while her husband, James, worked on a cattle ranch. Judy and the kids got busy building a fence with some big timbers, and they all had to pitch in to move them into place, figure things out, and cooperate. It took a long time, but they got it done, and when James came home, Judy was very proud of their accomplishment.

James took a look at the fence and then went to one end and looked down the fence line. Judy was surprised. "He asked me if I had eyeballed it. I didn't know what he meant, and then he showed me. Our fence was all crooked." It was not something they had paid attention to, and she didn't realize she even needed to.

"I got it done, the animals were in—shouldn't he have been happy?" Judy continued, "It didn't matter to me, but he liked precision and wanted straight lines. It mattered to him because it was a reflection of his place. We learned what's important to the other person by working back and forth with each other."

Families should check their expectations that the children will carry on the farming torch. For Judy's half-dozen, who grew up with goats, horses, chickens, and a garden, the results are mixed. Her daughter Elizabeth thinks goats are just a problem, but she has always had a passion for horses that she is now turning into a business.

Melissa does it all on her farm.

Judy remembers, "That was something I originally wanted to do, but it wasn't so much for me because there was so much to learn to get there. She's had a start with it that I didn't have, and she's got a natural talent for it." Judy admires how Elizabeth has taken her passion and honed it through dedicated training.

Judy's other daughter, Rachael, doesn't see horses or any farm animals as necessities, but she enjoys them casually and recognizes how enriching they've been in her life. If her kids want them, then she'll accommodate them, but for pleasure only. One of Rachael's daughters already displays a deep affection for horses and by age two was spending a great deal of time interacting with horses by her grandfather's side and riding with her Aunt Elizabeth's guidance.

Judy's sons also developed distinct opinions about having livestock. Benjamin and Nathaniel both would like to have animals if someone else takes care of them. But Judy was surprised to learn recently that Samuel, a burly yet sensitive welder, loves chickens.

A few of Aunt Judy's herd.

"He likes just having them around. He likes their noise, watching the way they scavenge, and it just soothes him."

Daniel, Judy's first child, was the reason she became a goat farmer. He had an allergy to cow's milk, so the family who had never farmed before got goats. Not all parents would take that leap to provide fresh milk for their babies. As a military man in his forties now, Daniel still craves goat's milk. Judy shared this anecdote to describe the lengths he would go to: "Daniel was overseas in Hungary, and he could not check out a vehicle but could check out a bicycle. So he got a bike and rode a few miles down the road to find a goat dairy to get the resource that he valued."

For this family, the entertainment value outweighed the feed, vet, and fencing costs of having horses and other animals. Judy recalls, "When they were kids, when we watched and observed the way they interacted with horses, we knew they needed it. Against everybody saying it's too expensive." Judy and James saw beyond that. "There was something they needed from that environment—responsibility, but not only that—a place to interact, create a relationship with responsibility, and enjoyment."

Look back at the list of your main wishes and desires. Have your partner complete the same exercise. Compare your list with your partner's. What overlaps? Where are there points of conflict or concern?

Work together to create a new list of five main desires for your relationship.

Now, how do these words connect to farming? Those qualities that you hunger for—which of them can be satiated through working with the land, raising animals, harvesting wild herbs, or canning and preserving? In what ways will building a greenhouse or planting a food forest bring you closer to your dreams? Brainstorm about each of your five main wishes and how they tie into your ideal farming lifestyle.

Lisa shares the fruits (and vegetables) of her labor at farmers' markets.

EXPRESSING YOUR VALUES

Alvina Maynard loves the farming lifestyle for her young children but is grateful that her farm lies just minutes from an interstate that connects her to the city. Her alpaca farm is not simply a space for her kids to enjoy a farming lifestyle—it incorporates her mission to change the way consumers affect the textile industry. When I visited her farm, she was conducting a tour for college students majoring in product development and fashion. Alvina has done her homework, and she strongly believes that she can make a difference. "You have a 300-percent increase in textile waste since the 1980s. Three hundred percent that's going into landfills. Most of that is synthetic fiber. You all are going into fields that are going to influence change so that we can have more sustainable textiles," she told the wide-eyed students on the tour.

Even for educated and environmentally conscious consumers, we are much less aware of the environmental impact, the sustainability, and the human-rights issues that touch our skin through our wardrobe. "The timing is right in terms of natural fibers. We're paying more attention to where food comes from, so now I can leverage that traction: where do your clothes come from?" Alvina's question is difficult to answer, but she doesn't shy away from difficult conversations. She believes in bringing this conversation to the forefront, to the general consumer, so that we realize the impact of our dollars, which can either support a sustainable, healthy future or support the continuation of damaging the earth and its inhabitants.

Lisa Munniksma's values are reflected in working with animals. "Livestock farming is important because I don't eat meat unless I know where it came from. I do believe animals are here for a purpose but also that they should live the best lives they can possibly live while they are fulfilling their purpose. I realize that people are not going to stop eating meat. But I want to give people the option of eating meat that comes from animals that are truly cared for and that live sustainably."

Sometimes the longing to work on a farm is there, but we don't know why. We can't foresee where the path will lead us. While Lisa had a history of enjoying horses and working with them, it wasn't until she committed to owning one that her normal life became a farm life. It began with a passion that surpassed logic and led to a deeper understanding of what it would require of her. "I loved horses first and then figured out how to manage the equipment I needed. I got a horse, so then I got a truck and trailer. I couldn't rely on other people to drive me around, and I had to learn to be self-reliant."

Lisa trained with a couple who she credits with changing her life as much as her decision (years later) to leave her full-time job to travel and farm. They helped her see the therapeutic qualities of working with an animal. Lisa admits that she had a "problem" horse. "It turned out that I was a 'problem' person and would have had the same problems if I got another horse today."

KNOW

What do you know?

Quickly jot down as many things as you can think of about your own knowledge of farming.

In preparation for talking with the women I interviewed for this book, I made a list of questions that would help me get a baseline understanding of their experience in the field. You can use the same questions in your self-assessment. Answer the following as thoroughly as possible.

What is your primary role or title?

Other roles or titles?

How long have you been in your primary role?

Who else farms/gardens with you?

Where and how did you learn your skills? Please describe briefly.

Did you grow up in a farming family?

Formal education:

Apprenticeships or internships:

Mentors:

Self-taught skills:

Other (conferences, trainings):

What resources do you wish you had when you first began farming?

What resources do you wish you had once you'd gained some experience but still needed help?

What resources do you still need today?

CONFERENCES

Every season brings its gifts, and with winter comes farm conferences. When the fields are put to bed, learning and networking opportunities blossom across the United States. Conferences are efficient ways to learn new skills, explore potential markets, network with the farming community, and pick up new ideas. You can attend the EcoFarm Conference in California, the Midwest Organic and Sustainable Education Service (MOSES) Conference in Wisconsin, the Northern Michigan Small Farm Conference, the North American Biodynamic Conference, or the Southern Sustainable Agriculture Working Group (SSAWG) Conference in Kentucky, to name just a few.

Getting away from the farm for even a few days can be tricky. You may need to find responsible animal caretakers, justify travel expenses, buy tickets, and pay for hotels. All of that trouble, and it's not even a vacation! Or is it? If you work it right, you can have fun, learn, meet new people, and experience a different region's culture, which adds up to a worthwhile investment in your farm business.

Conference such as Southeast Wise Women offer a supportive, female-only environment.

Here are some ways to make the most of a conference before, during, and after the trip. If you read no further, this one simple tip will pay off: take notes.

Before the Conference

Be purposeful. Think about what you are doing well and what you really want to learn. Consider the past year's successes and write down a few main trouble spots. Consider whether these issues are geographical, biological, philosophical, or business-related. Consider why you are farming and what kind of moral support will boost you. Contemplate ways in which you'd like to challenge yourself and ways in which you hope to grow.

Choose wisely. Look through the conference schedule, including pre- and post-conference workshops or field trips. Consider whether a single day would cover your interests or if it would be worth the money to register for multiple days. Is there a cancellation policy? Is transportation offered for off-site sessions? Are hotel rooms available at discounted rates for conference attendees and, if so, is there a deadline to make reservations?

Consider whether you want to bring along the family or if you'd rather embark on a solo personal-development journey. If you want to visit family or friends and make a vacation out of it, look for good workshops or conferences in their area and then plan to stay a few days longer for socializing and sightseeing.

If you are staying overnight, you might want to treat yourself to a Jacuzzi and room service in the hotel. Or you could round out your farm-networking experience by staying at a nearby farm bed-and-breakfast, where you could learn a thing or two. Check out www.farmstayus.com.

During the Conference

Come prepared. Presenters appreciate when attendees ask thought-provoking questions. If you have your list of questions and you know what you came for, you'll be a good student. Balance your main goal with curiosity and imagination. Find something you've always been interested in but have never tried, and attend a session on that topic with an open mind.

Be flexible. Sometimes golden moments happen between formal sessions, when folks are milling around, chatting, and the pressure is off. Presenters can speak candidly over lunch or at the bar, and you could get personalized answers and meaningful conversations with both the presenters and the attendees who represent different perspectives.

After the Conference

Debrief and share. You may feel overwhelmed with all of the new information and ideas you come away with. Before returning to your normal schedule, take the time to write up your notes. Summarize ideas and organize papers so you can find them again. Hold a short meeting with your family or workers to discuss what you've learned.

Look back at the initial questions you prepared for the conference, and choose one of the things that you learned. Make a plan to apply that skill or concept to your farm practice for the next six months, with weekly check-ins. Hold yourself accountable to just that one thing and take note of what happens.

DIFFERENT PATHS TO KNOWLEDGE

Knowledge seems to be ephemeral and intangible, with no real way to measure it until it is put to the test. It's always relative, and it's never complete. Take a look at the diverse routes some women farmers have taken to become educated and knowledgeable about their farming practices.

Jessica Ballard did not grow up around farming but was drawn to culinary school to learn about preparing healthy food. That led her to what she calls "making the hippie

WHAT IS PERMACULTURE?

Permaculture is a holistic design system based on natural ecosystems. The word permaculture indicates a permanence in culture, and it is most often applied specifically to agricultural systems. However, the ethics of permaculture—earth care, people care, and fair share—encompass much more than growing food sustainably.

Earth care focuses on observing ecological functions and recognizing our place in the system as we design gardens, farms, organizations, or households that work with the flow of energy and resources. People care refers to a basic need to sustain humanity, whether in the context of a global need to feed the world or as local priorities, such as paying a fair wage and creating work environments that support people without harming the earth. Fair share is also called return the surplus, meaning that we don't take more than we need from the productive systems, and we limit consumption while distributing the surplus equitably. In permaculture, we reinvest our profits, whether it's through composting food scraps, saving seeds, or paying employees. The purpose of a permaculture system is to proactively regenerate, repair, and strengthen the earth, just as the earth provides for our needs. Woven throughout the three ethics are intertwining philosophies of social justice, human rights, and environmental stewardship, and twelve guiding principles shape the permaculturalist's approach to designing a system.

1. Observe and interact: Before you can begin to modify your environment, you must understand its patterns and possibilities and be able to watch and work with nature's cues. Return to this primary principle again and again.

2. Catch and store energy: The cycles of nature rule our bodies, our days, our animals, and our plants. Notice when resources and by-products are most abundant and plan your action to maximize their potential.

3. Obtain a yield: Our systems must be reciprocal to be sustainable. A yield is not limited to a monetary profit or the fruit that a tree bears. Yields could be knowledge gained from firsthand experience, physical renewal from outdoor work, improved water and air quality from green plants, and community consciousness through collaborative work.

4. Self-regulate and accept feedback: Solutions arise organically out of problems if you take the time and interest to understand the situation. Sense what is out of harmony and recognize your own role.

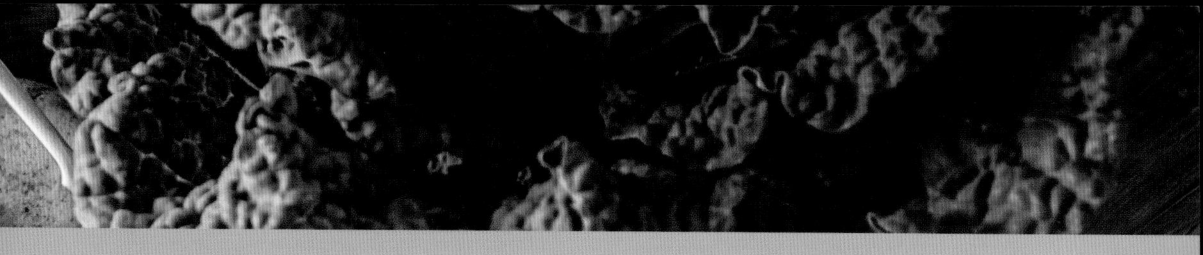

Take your cues from nature's tendency to create balance, and modify your behaviors to enhance homeostasis.

5. Use and value nature's gifts: The services nature provides are beyond measure and can strengthen a system on many levels. Learn about the most abused and least appreciated ecosystems in your area and restore a portion of your property. Imagine relying on animal power rather than fossil fuels. Learn how to responsibly utilize potentially renewable resources (sun, wind, hydro) as your system's allies.

6. Create no waste: Nature has no discards; everything is of value. Compost, recycle, reuse, repurpose, and upcycle everything you use. Reduce or eliminate anything that cannot be "digested" by your system. If you can't use something in your system, connect with another system that can.

7. Design from patterns to details: Energy efficiency and productive functions result from well-designed forms. Branching, spiraling, and triangular patterns show up in nature everywhere. Cultural patterns influence and inspire our systems as well.

8. Integrate, don't segregate: The whole is greater than the sum of its parts. Inclusion of diverse people strengthens social dynamics in a community garden. The elements of a homestead can support each other if each part serves many purposes.

9. Small and slow solutions: Slow food, slow money, and slow fashion are examples of movements embodying the idea that a healthy lifestyle is dependent on making clear-headed, premeditated decisions that will affect change in incremental, yet powerful, ways.

10. Value diversity: Diversity is the backup plan to the backup plan, or intentional redundancy. We can expand the limits of our systems by integrating a variety of plants, animals, structures, and ideas that bring unique attributes and overlap functions.

11. Mind the margins: Action begins on the fringe of a system, which often supports the most vibrancy and diversity of life. Permaculture has arisen as a marginal concept from a variety of mainstream disciplines, including philosophy, architecture, landscape design, forestry, agriculture, and economic development.

12. Creatively respond to change: We are cultivators, indicators, observers, and participants with nature and the systems we design that mimic and maximize nature's gifts. When we are in sync with our vision, our space, and our community, we can intervene at the right time as we refine our creative course.

Permaculture is founded on respect and reverence for the soil, air, water, and plants that our ancestors formed relationships with from the beginning of humanity. Its goal is to utilize the tools we modern humans have not only to simply sustain our current level of consumption but also to enjoy a lifestyle of fruitful and mindful prosperity that restores and regenerates the health and stability of all life.

THE RING OF KNOWLEDGE

The following is a modified version of a post the author wrote for the Southern Sustainable Agriculture Working Group's blog, SSAWG BLAWG, www.ssawg.org/ssawg-blawg, published on March 15, 2016.

I don't expect to feed the world or even fill a shelf in a grocery store. I do expect to grow most of my own food and share a surplus. My plan is to steadily increase my skills and growing space until I have a full larder consisting largely of homegrown food. I'm sure I can do it. I want to do it. Do I know how to do it? Well, yes and no. That's why I went to the 2016 SSAWG conference—to learn enough to feed myself and share the surplus of information with readers.

In my blog posts, I'll recap what I learned at the SSAWG conference, where a diverse field of topics enticed farmers of all levels of experience to enrich their knowledge, hone their skills, and boost their motivation to farm for the greater good.

On the first morning of SSAWG conference sessions, Joel Salatin drew a big crowd. I, on the other hand, drew circles. Three rings, overlapping in the middle. This simple Venn diagram would become a symbol of my farming future.

"If time and money were not an issue, what would you do tomorrow?" Joel challenged his audience with this question, and I followed his instructions. I wrote the words *love* inside one circle, *good at* inside another, and *know* inside the third.

Just because I know a lot about something doesn't mean I'm good at it. Just because I'm good at something doesn't mean I love to do it. And so forth. Joel's goal with this exercise was to encourage movement toward the center of the diagram, where the rings overlap. This simple reminder becomes a compass, guiding me toward the sweet spot in the middle of those three realms. The focus area is the place where what I love, what I'm good at, and what I know how to do all come together.

My current relationship with farming falls outside of the center of the three rings. I know a lot about organic farming, and I love to work outdoors with plants and animals, but I have far less hands-on experience with growing food successfully.

I was an armchair farmer in the high desert and dusty plains of the West, reading Barbara Kingsolver's *Animal, Vegetable, Miracle* while daydreaming about rain. I made a deliberate move to settle down in Kentucky, a place where green things actually grow pretty readily. I was a short-term volunteer on a few farms and spent two years helping organize and tend community gardens. Then I tried to lead middle-school students in creating a permaculture urban farm. That's when I realized how little I actually knew.

Maybe it's true that those who can, do, and those who can't, teach. Well, I'd like to become a better do-er. To fit "teach gardening skills" into my three circles, it would have to go inside the "know" circle. I know how to do it but am not altogether good at it. While I love teaching gardening sometimes, I truly enjoy the act of working alongside colleagues of all ages and sharing experiences. I heard about SSAWG and wanted to absorb more knowledge so I could practice more in the field, on my own at first and potentially with others.

Sessions at the SSAWG conference could fill up my "know" circle. Everyone there was an educator in some respect. Expert farmers shared a wealth of knowledge in the form of practical, useful information. I will definitely refer back to the recommendations from the following presentations.

"Intensive Vegetable Production on a Small Scale," Pam Dawling: Intensive is right. While Pam's title referred to growing a large amount of food on a small amount of land, her presentation was like a graduate-level course jam-packed into seventy-five minutes. The great thing was that I didn't take any notes and could absorb her wisdom. She gave a handout with very detailed examples of transplant age and size for a sampling of veggies and spacing for crops for various goals (early harvest, maximum yield, sizes, etc.). It also summarized biointensive integrated pest management, sustainable weed management, and season extension. Pam's presentation, which is one of her many presentations available on SlideShare.net (search "Pam Dawling"), went into great detail on the timing and rotation of crops specifically for her ecoregion. Her book *Sustainable Market Farming* is rich with tables and timing plans to get the most out of a small area.

"Permaculture Designs for Small Farms," Shawn Jadrnicek: This was another one that left my head spinning, partly wondering why this guy hasn't been recruited by NASA to establish a self-contained life system on another planet. I think he's got what it takes. While I have a decent understanding of permaculture practices, I'm continually amazed at the ways people like Shawn can work with nature to enhance growing systems.

Much of his presentation focused on a pond. What does that have to do with farming? Everything about his pond is hosting a life form, impacting a microclimate, building soil, and performing at least ten other functions. This qualifies as biointegration. The system is alive. If you're more into chickens than watercress, his flower-petal design for rotating movable fences just outside his door is impressive—simply elegant and highly efficient. Read all about it in Shawn's book *The Bio-Integrated Farm*.

One of the biggest benefits from the conference was the abundance of takeaways, literally: handouts, resources, literature, and references on best practices for everything from when to plant to how to sell. SSAWG could compile a thorough sustainable farming library just from the websites and books the session presenters authored.

The sessions I mentioned were ones that I classified as fitting into my "know" ring. They provided statistics, percentages, ratios, specific crops, and timetables, all of which can be great starting points and troubleshooting guides for my future farm. I may be too green of a greenhorn to really know what good all this accounting, science, and engineering will do for my little hobby farm. I don't want an organic chemistry course to be a prerequisite for planting a potato in my backyard. Thank goodness I don't have to remember it all now. I don't even have to understand it all now. However, maybe I should track my sprouting sunflower seeds on a planner. It's never too early to start a good habit, and maybe I could sell sprouts in the future. You never know!

GreenHouse17 uses farming as therapy for domestic-violence survivors.

circuit"—traveling around and helping out on farms. She did an internship on a farm, and then another, and then finally decided to pursue farming seriously. She earned her bachelor's degree in sustainable agriculture at the University of Kentucky, and she proudly rejoices that she graduated debt-free as a single mom.

She got involved in local community gardens, which led to a position at Bluegrass Domestic Violence Program, which has rebranded itself as GreenHouse17. Over six years, with Jessica's influence and persistence, the shelter's farming program has grown to become the symbol for the type of care the organization provides for people going through crises in their intimate relationships.

Melissa Calhoun cannot simply list one job title or role because she is multitalented and does not restrict herself to one identity. Like Jessica, she was new to farming in her twenties, and she was also a bit disillusioned by her college education. "I went to school to learn how to save the world, basically. Studied ecology and environmental science. The political end was depressing. It was in Atlanta, and everything was growing up. Developers win every time. It's like the study of destroyed environments. Urban ecology was interesting but not necessarily inspiring my heart, I guess."

Melissa needed real-world, practical understanding of the concepts she had studied. She worked on a stream survey, which, she says, was the first time she had connected with the environment in a vast way—not just the trees in her neighborhood. "Being in the Appalachian forest felt like home. All the other people would go home, and I'd have the whole place to myself. I heard deer snorting for the first time, and I was like, 'What is that?'"

SHARE!

Some conferences offer an online forum to connect potential rideshare and roomshare partners. Find a room to rent from locals at www.couchsurfing.com or www.airbnb.com.

Melissa readily admits, "After I graduated college, I felt I had no skills." To remedy this, she found opportunities to exchange work for rent as she learned how to work with her hands. At a permaculture apprenticeship in North Carolina, she learned how to make a hoop house out of bamboo, create contoured beds, and plant lots of trees. At Sequatchee Valley Institute in Tennessee, she immersed herself in a permaculture library, learned about plants and natural building, and taught others through giving tours and training interns.

She and her husband got firsthand experience with a homestead startup they found by putting up flyers that said, "We want to help you farm or landscape." A family had them over for dinner, and they agreed to work together. That led to more hands-on natural building experience, including making straw-clay slip bricks in August. Far from her college knowledge, Melissa says, "That's some of the hardest work, next to haying in Maine, that I've ever done."

Delia Scott's path has been more straightforward, but she never intended to become a farmer. Instead, she has provided countless hours educating others on sustainable growing. Her bachelor's degree in horticulture and master's degree in organic and sustainable agriculture prepared her, but it wasn't until she had internships and jobs at nurseries that she confirmed she really wanted to do this work. She moved into a position as a horticulture extension agent for her county's cooperative extension. By working with community gardens, schools, the Farm to School initiative, and Master Gardeners, Delia developed her abilities to engage the public, which led to the role she now fills.

Delia is the first executive director of the Organic Association of Kentucky, a nonprofit dedicated to improving the health of people and the environment by educating and providing resources for farmers and consumers interested in driving change toward a more sustainable food system. They hold an annual conference that attracts the greatest minds in sustainable and small-scale farming and features esteemed keynote speakers, such as Wendell Berry.

Delia Scott works toward more sustainable farming in Kentucky.

HAVE

What do you have?

Quickly list the assets you have that can help you in your farming project. What are your strengths and readily available resources?

When we confront a new project, we often go to extremes. We may get super fired-up and passionate and blindly forge ahead, or we may see how the odds are stacked against us and allow the obstacles to overcome our desires. This section will help you take some real action steps, all still in the planning phase, to assess your resources. Taking these steps will help you in doing the type of farming you want to do and will help you identify your limiting factors, including time, finances, support from external resources, physical ability, access to land, and imagination. A lack in any of these areas does not and should not mean "stop." Rather, find your surplus and see if you can leverage that to help you further your goals. Let's get started.

No matter what your farm dreams consist of and how far-fetched they may seem in your current situation, you may find that by assessing your resources, you have more than you realized. Read over the list you just created and notice how you defined assets, strengths, and resources for yourself. As you move through the rest of this section, you may find that your perception of an asset will shift. Keep an open mind.

Barbara Stanny's book *Sacred Success* provides an in-depth exploration of the stages of pursuing a dream, beginning with a rite of passage centered on surrender. She offers this wisdom to ground and support you when stepping onto the new path: "Where you are right now in your life, whatever is going on, is perfect. Where and how far you go next depends on how much you're willing to release and what you're willing to receive."

Assessing your space allows you to make the best use of it.

Starting with where you are right now, whether it's the property where you live or a weekend getaway, a piece of land you own or the apartment you rent in the city, take stock of all its characteristics. This practice will not only give you an inventory of what you currently have to work with, it also connects you to your space and builds a deep relationship with it. Try to incorporate a mindset of observing patterns, noticing where things are integrated, and cultivating your vision along with nature's cycles. The following exercises are more than just planning steps. Prepare to make them lifelong habits, revisiting and refreshing your view of your space when you sense a need for a new perspective.

Everything moves, flows, cycles, and transforms. Energy, electricity, fluids, money, air, blood. This theme unites the following exercises of assessing your property, your finances, your physical abilities, and your external resources.

YOUR LAND

Farming is about connecting with the land, but the land is so much more than the dirt beneath your feet. Your property is a multidimensional space. You can play with it in so many ways! The first step in assessing your land is mapping it.

Look to permaculture for an all-inclusive philosophy and approach to understanding resources and the way they transform, add value, become depleted—in short, how resources flow. Before tracking the flow of each resource and how it affects other areas, let's first map out each layer of your property—not only the seen but also the unseen, such as wind. The goal of this map is to incorporate the permanent, semipermanent, transitory, and ephemeral, and put them all together in one place. Each layer of your map will show a distinct element, a part of the whole.

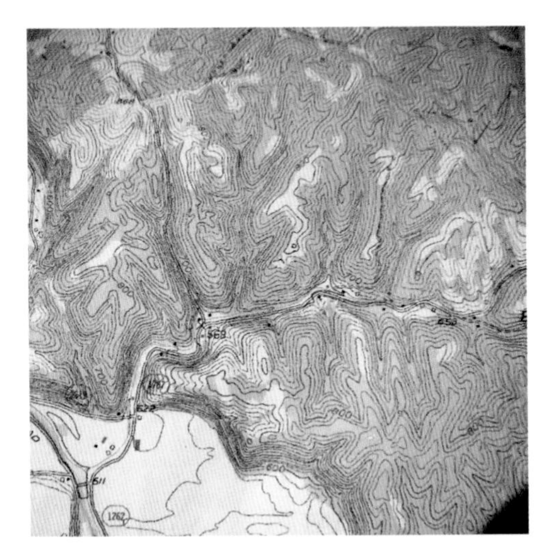

Mapping is a fun, multisensory exercise that you can take as far as you want. For starters, simply follow the instructions and encourage your family (and farm team, if applicable) to be involved. Separating the layers and allowing team members to work independently and report back at the end is a great way to stimulate conversation and explore unusual dimensions of your landscape.

Flows

Don't be overwhelmed by it all, just go with the flow. The easiest example of a resource flow is water, which usually moves from uphill to downhill. But the flow story doesn't end there. Water can evaporate, precipitate, and fall again, or it can pool, whether for a few hours as a puddle after a rainstorm or indefinitely as a stagnant pond. The important thing is to pay attention to where water is flowing on your property and where it is not.

Can you imagine the same idea applied to another elemental resource, like wind? Air definitely flows, and you may experience a prevailing wind from different directions at different times of year. Your farm's structures could be shaping the air's flow, funneling wind into a forceful channel or pleasantly protecting plants or animals.

Now think of an animal resource, such as chickens. They provide compost, a resource that flows from the chickens to the garden. The garden may also be a resource for the chickens. Plus, the eggs from the henhouse will flow to your kitchen or to the neighbors or to the market.

To visualize the flow of these materials and potential energy, go to your layers map and draw arrows between areas to indicate movement. After you have identified all the ways that your property is feeding itself, recycling its resources, and generating its own energy, pat yourself on the back and enjoy the self-sufficiency you have achieved. Then look at the potential for closing the loop on your own property so that you are providing more of your own resources and nothing is wasted.

You can take this exercise into a bigger realm and expand your flow map to include your watershed, your neighbors, and what is upwind or downstream from your farm. Pollinators fly over fences, and groundhogs burrow under them; likewise, your property lines do not halt resource flows. Use this mapping technique to make connections in your own mind about how your farm's healthy productivity connects to different points of origin and get a sense of the ripple effects that your farm passes along.

From what you've learned from the mapping exercise, jot down your thoughts on how your farm impacts the community.

Aunt Judy tells us to observe. She warns, "You can't take care of illness and see a problem coming up if you aren't a good observer." Her experience on a southern Texas cattle ranch put her observation skills to the test, and it became a tool by which she measured farmhands she might hire. She and her husband, James, were employed to keep an eye on sheep and goats in huge, five-section pastures (each section was 640 acres). She recalls, "They turned the sheep out, and there were huge ravines, and we couldn't find any. We went back and forth looking for weeks, and we told the boss they weren't there. One day I caught something. I'd been looking at the same terrain, and I began to see them. James and I learned that when we hired people to help us, we wanted to find the people who observed and saw. You can't be a husbandman if you can't see. Anybody can run livestock into a pen, but the ones who are going to save your herd are the observant ones."

Close observation doesn't come naturally to most of us. You have to train yourself; nobody else can do it for you. Judy is glad that she raised her kids and has run her childcare business on the farm. She says, "A lot of farm kids are natural observers and they don't [even] know it; they've just grown up seeing things."

Observation is the number-one principle in practicing permaculture. Judy has never studied permaculture, but, like many natural farmers who learn by watching, she has innately understood the importance of the observer: "If a person learns to be observant, [he or she will] see the natural law at work. We use technical advancements, but…in harmony with what you know. See what works and incorporate it with what works."

MAPPING YOUR PROPERTY'S LAYERS

Supplies

- directional compass
- grid paper
- transparency sheets, minimum five
- dry-erase markers, minimum five different colors
- clipboard or other firm surface
- binder clips

Instructions

1. Secure a piece of grid paper to your clipboard as the base layer and place one transparency on top. Secure it with a binder clip so it doesn't slide.

2. Establish the base layer. Use the compass to find true north and mark it on the transparency. Use the grid to measure your property to scale (optional) or draw freehand the edges that bound your land. Draw the boundaries of your property with one color on the transparency. Use the same color to add existing structures (fences, home, barn, outbuildings, and impermeable surfaces, such as driveways). Use symbols as needed. Keep it simple.

3. Add another transparency and secure it. Continue to include as many of the following layers as possible, one at a time, using a different color marker for each, if possible. Start with what seems easiest for you. Feel free to skip any that don't make sense right now. Make a point of revisiting the more difficult ones as you progress.

   - underground features, if known (pipes, wiring, septic tank, karst features)
   - vegetation (designate as perennial, annual, cultivated, wild, ornamental)
   - soil types
   - corridors of travel (human, automobile, livestock, wildlife, pollinators—consider airspace and waterways as well as ground)
   - summer sun and winter sun (note position at regular times of day as well as areas of constant sun or shade)
   - wind (notice this in as many seasons as possible; be aware of leaning trees and odors carried by the wind)
   - topography (changes in elevation)
   - water sources and direction of movement (a rolling tennis ball can help you see slopes on smooth terrain; where it stops, water is likely to pool)
   - wildlife (areas providing habitat of any kind: food, shelter, water, space)
   - sounds (mark where sounds are heard, rather than the source, e.g., traffic, birds, wind; notice changes near structures)
   - pollution (in any form, such as trash from passing cars, noise from sirens, odors from neighboring farms)
   - cultural features (old foundation, springbox, spaces of significance, sites of important memories, favorite relaxation spots)
   - views (landscape, neighbors, windows, perspective points)
   - risks and hazards (fire danger, open pits, dangling tree limbs)
   - temperature zones (cold sink, thermal mass)

Judy shared another great example of how children and adults can learn to ease their efforts if they watch animals at work. As a child, she bravely took her uncle's draft horse out on rides, often bareback. "We found that when you want to cross a creek, in Iowa, the creeks aren't necessarily little, but they're like bogs. Our horses wouldn't go in, and we'd get aggravated and switch them a little. Then the horses would start to cross and then jump! I don't know how many times we ended up stuck in the bog, and we learned why they didn't want to cross the creek at that place. So we found that up and down the creek, there were crossing places. We watched the cows, and where they would cross, it was all packed. You see what we did? We were learning to observe. We tried it our way, and it didn't work."

Taking Inventory

Again, just observe—with all your senses. The longer and more frequently you do this, the better off your farm will be, and I could say the same for your peace of mind. You will be more in tune with the subtle changes and little quirks of your land. Big events won't catch you off guard because you will have an intimate sense of all that is occurring around the farm or garden.

If you don't live on your farm, take at least fifteen minutes every time you arrive to walk around and use your senses to notice, without taking any action at all (barring emergencies, of course). Use wide-angle vision. Expand that to your other senses.

Evaluate your property and create what you envision.

You are an artist. You can create anything you envision. Understand the canvas, the paints, and the brushes, and listen to your muse. Collaborate with nature and other people. Encourage input from visitors to your property. Hand them a notebook or clipboard to make observations as they spend time with the land.

This will lead you to a satisfying place of seeing what is worth keeping, what invites transformation, and what is ready to be let go. If you are in a phase of renovation or construction, Peter Bane suggests in *The Handbook of Permaculture* what to do at this point: "Think about how you can enhance the beneficial influences and temper, block or mitigate the hostile ones by how you place and build things on the farm."

If you are simply getting a feel for the space you inhabit, dreaming of the garden you'll start, or figuring out how to shift some existing systems, spending time on observation will give you a better idea of the inventory you have to start with.

A farm of any size can offer opportunities to engage all who visit it by sharing the role of observer. The more people involved at all times of year, the fuller understanding you can develop of the place. A few ideas for capturing data include:

- Keeping a log book in a central location for guests and residents to make simple notes
- Providing a checklist of wildlife on-site with space to add new species
- Creating a simple note-card template with observation points for guests to record, such as wind direction, weather conditions, and insects and/or plants identified. Keep these cards organized in an index-card box, sorted by year, season, or location. Add photos when possible.
- Beginning a journal, almanac, or album, which is intended to grow as knowledge accumulates, to pass on to further generations. It's OK if you don't know much right now, or even if you plan to move. Just starting a page to add to later will encourage and remind you of the community you share and the legacy you'll pass on. Think about who lived here before you and before them. How far back can you trace the lineage?

Here are some ideas to get you started on an inventory of your property:

- Create a list of plants. Feel free to name them in your own way if you don't know what they are; don't get hung up on classifying them yet. This stage allows innocence and imagination. Connect with the plants in whatever way works for you. Add new plants as you discover them.
- Imagine you are creating a field guide to your property. Check out local field guides from your library and decide on a format you'll use for your own. Include drawings, photos, leaf rubbings, or small plant samples.

USE EXISTING RESOURCES

Your planning should be informed by maps that already exist. Make it a priority to acquire these for yourself, whether you rent or own your property:

- plat (city or county) with property boundaries
- topographic map (www.usgs.gov)
- aerial view (www.google.com/earth)
- soil map (websoilsurvey.sc.egov.usda.gov)

- Begin your foray into phenology, making observations that correlate biological occurrences with the climate, such as:
  * Buds of flowers: When they appear and open up
  * Leaves: When they first appear
  * Seeds: When they are ready to be collected
  * Berries: Whether edible or not and when to harvest
  * Roots: If edible, when to harvest
  * Fungi: Edible or not
- Create your property's list of animals. Remember to start small—very small—with insects. You'll learn an immense amount about the health of your plants if you pay attention to which animals are working on your farm. For a fuller exploration of the wild animals who share our space, check out the book I coauthored and edited, *Wildlife in Your Garden* (Lumina Media, 2016). Using the same idea as the field guide you are making for your plants, include a section for each of these categories:
  * Insects: What plants or habitats you associate them with
  * Birds: When and where you see them and songs to listen for
  * Fish, amphibians, reptiles: How they behave and what they eat
  * Mammals: Small to large, observe before you react

Using your imagination and making up names can be a fun and lightweight introduction to your space, even if you have lived there for generations. Many people have no idea what the names of the plants underfoot are, much less if they are useful to us. To really understand your place and all of the life within it, you may eventually train yourself and become an amateur naturalist. Or, if you want to speed up the process, make the acquaintance of those who specialize in studying the natural world of your

When it comes to your land and your living arrangements, there are many options.

region. Consider inviting professionals, such as foresters, herbalists, native plant gurus, or university staff to your property as well as hosting groups or researchers who want to study something that could be unique to your site.

Ownership or Alternatives?

Land seems like a simple prerequisite for growing food, fiber, or medicine, but you don't have to buy in order to grow. There are so many creative ways to approach farming that the classic farm-inheritance scenario seems like a relic of a bygone era. Urban farming on blighted properties, agritourism as part of farm-to-table restaurants, harvesting wild herbs on other farmers' marginal land, providing horticultural therapy by gardening with social-service agencies, growing food in trade for rent—the possibilities are endless, all without taking out a loan to buy a big piece of acreage.

Many people who dream of becoming farmers are stopped in their tracks by the heavy cost of land, but that doesn't mean they need to give up altogether. You don't have to own land to be a farmer, but you do need access to it. There's no single path to finding a little slice of farming heaven. When it comes to stewarding plants and animals, an open mind and adaptability—not real estate—are your most valuable commodities. Here are some real-life examples of ways to think outside the ownership box.

Renting Versus Buying

One advantage of not owning land is the flexibility to try out farming enterprises without the added costs of a mortgage and being tied to a single place. If you don't necessarily want to live on the land you work, renting is a good option.

When it comes to land prices, horticultural land is more expensive than pasture or hay ground, although you don't need as much of the former. Finding the right land in the right place is the main challenge, and you should be prepared to compromise. When you make your farm plan, Dr. Lee Meyer, extension professor in the Department of Agricultural Economics at the University of Kentucky, suggests considering four key questions because knowing what you need and what you can afford is the starting point.

- How many acres do you need?
- What are your goals and time span?
- How important is flexibility?
- What is your financial situation, both in cash flow and equity?

Tending with Neighbors

Community gardens are popping up all over urban and suburban areas, run by nonprofits, churches, neighborhood groups, and other organizations. Volunteering at or renting a plot of land in a community garden can be the perfect entry point to trying out your green thumb before you take the plunge and acquire your own land. Working alongside other gardeners of varying experience, you'll gain a wealth of agricultural knowledge without taking a class or running a tractor. Best of all, the gardening activities bring new life to areas that may have been blighted or neglected.

Seedleaf (www.seedleaf.org), a community gardening nonprofit in Lexington, Kentucky, offers training for Master Community Gardener certification, which is similar to the cooperative extension's Master Gardener program. In the series of evening and weekend classes, students learn how to plant, cultivate, and harvest not only food but also connections in the community. Gardens serve as welcoming spaces to host neighborhood barbecues, bring children into nature, and encourage beneficial wildlife. Food is offered freely to all who enter the gardens, and cooking classes teach the benefits and sensory excitement of preparing fresh vegetables. If you

Community gardening can be a perfect venture for aspiring hobby farmers or those in urban areas.

want to try out a specialty crop, experiment with growing food, or just get ideas, step into your nearest community garden.

Whether you live in the city or the country, remain open to sharing with neighbors. Paige and Eric Quillen own five rural acres in Versailles, Kentucky. When they moved there in 2001, they thought their hobby farm would offer plenty of space for a few horses, some sheep and chickens, a dog, and a few cats. The five acres turned out to be not quite large enough. Fortunately, an elderly neighbor who shared a common fence with Paige and Eric expressed his wish to no longer mow his back two acres. After sitting down to talk through the possible options, all parties came to an agreement: the Quillens paid for new gates and fencing, and their neighbor allows their horses to graze on his land. Now the horses are happy to have more pasture, and the neighbor is happy that he doesn't have to mow.

Working for Land

Apprenticeships and work-trade agreements can take all shapes and forms. Alice Melendez and her mother, Laura Freeman (the namesake of Laura's Lean Beef), are natural farmers near Winchester, Kentucky. They have taken the idea of sharecropping and given it a more benevolent twist. Like many rural homesteads, the family's farm includes unused land and lodging. As a way to share their knowledge and resources, as well as connect other homesteaders with each other, they founded Plowshares Community. Their educational initiatives include an apprenticeship program, but this one is a little different. Rather than simply laboring as farmhands in exchange for food and lodging, aspiring farmers are provided a space to try out their own ideas in the field. In addition, Alice and Laura provide seed funding to jump-start the enterprises.

Apprenticeships offer real-life experience and plenty of opportunities to learn.

Alice describes one such project: "Jerred Graham took us up on the apprentice farmer offer: we provided some start-up capital and a place to live and work through an agricultural enterprise of his design. We will work through business planning and production hurdles, and, when it starts to make money, we'll share the profit.... We've taught several people how to process chickens, developed a great non-GMO

(genetically modified organism) feed blend with ingredients off the farm, and built our market." Jerred also works for pay on the larger farm. Alice explains that the Plowshares projects may not be self-sustaining as a stand-alone enterprise yet, but with the added support from the larger farm, it is a great incubator for smaller experiments.

Building Community before Building a Home

Brian "Ziggy" Liloia grew up in the New Jersey suburbs and never dreamed of becoming a farmer. However, in college, he read books that inspired him to learn woodworking and natural building techniques. He lived at Dancing Rabbit Ecovillage in Missouri for seven years and spent his time there building homes, gardening, and raising chickens and ducks. With time, Ziggy's aspirations outgrew the community. He says, "I wanted access to more land to be able to create a more full-fledged homestead based on permaculture design principles and as a site for folks to learn about natural building, permaculture, and sustainable and regenerative living. "

A friendship between Ziggy and his partner, April Morales, and a Kentucky couple blossomed into a mutually beneficial arrangement that has led to his recent land purchase. Tim Hensley and Jane Post run Forest Retreats, a wildcrafting workshop and retreat center. They were interested in building a natural home on their wooded site and offered housing to Ziggy, April, and friend Jacob Graber in exchange for building the new house.

Tim and Jane's timely support enabled the Ziggy, April, and Jacob trio—also known as "The Year of Mud" (www.theyearofmud.com)—to spend a year leading straw-bale and cob building workshops while searching for the perfect property for themselves. This time also allowed them to integrate into the local community. Was there a formal agreement? Ziggy says, "We never drew up a formal contract but based our exchange on trust that we would all balance out without the use of money."

The Year of Mud found a piece of land to purchase and moved from Forest Retreats to their new home, which is not far away. They plan to expand the community of land-sharing. "We will no doubt offer internships and work exchanges in the future. We're also planning on incorporating at least two more people onto the land full-time as part of our vision of creating a small income- and resource-sharing community."

The Land under Your Feet

Whether you rent, dabble on a friend's property, experiment with the support of an established farm, or find an innovative way to secure your own property, the most important thing is that you feel at home there. Paige Quillen gives us her thoughts on returning home after being away from her hobby farm: "There's nothing like having your own plot, where you can go and open up the windows and smell the smells. I really miss that when we travel. Everything has a smell, and there are certain sounds that tell me where I am. You can tell when the seasons are changing and live a little more connected to the Earth."

Talking with women for this book, I found a wide assortment of attitudes toward owning land. Helen Terry owns a health retreat center on a ranch that is home to rescued donkeys and horses. A long-time city dweller, she never intended to buy the place. She and her husband, Joe, went to look at a lake house and stumbled across the ranch for sale. They had only fourteen days to close on the property and made a quick decision. Now, Helen is the self-proclaimed "Ranchess" at Soma Ranch in Montgomery, Texas.

Lizzy McNeil is a graduate student working toward a license in marriage and family counseling and equine therapy. She lives in Kentucky, where she says it takes two acres to support one horse. She could continue to board her horses on the two different farms she could inherit someday: twelve acres owned by her mother and forty acres owned by her father. However, her goal is to have a therapy ranch with plenty of room for horses, and she hopes to move out West, where Lizzy says it would take forty acres to support one horse because water and grass are much more scarce. So why go? Like the classic cowgirl, she yearns for those wide-open spaces. She got used to the western way of life while she worked on a ranch in Wyoming—a ranch that is now run by three sisters.

Angela Wartes-Kahl's marriage brought her into comanaging Common Treasury Farm in Oregon with her husband, Garth, and his farming partner, Allison, both of whom lived on the farm ten years before Angela arrived. Garth and Allison bought the farm in the 1990s with a group of friends who were full-time farmers. As needs changed, many moved on to other jobs.

Helen Terry's Soma Ranch.

Winters can be especially challenging for year-round farming in the Pacific Northwest, and the Common Treasury team has embraced the seasonal cycles of farm help. This cooperative farm hosts at least two interns and several WWOOFers each year, with between four and ten apprentices at any

given time. After one year, if an apprentice is serious enough, he or she can buy into the property with a four- to five-year contract. The cost is low, with the trade-off of being in a remote location, about an hour's drive from multiple markets.

Their goal is to nurture new farmers and grow the supportive organic community, Angela says. "We're trying to make sure to keep the price low enough that anybody who wants to be a farmer on their own land and grow their own food can do so without undue hardship—not having to work a huge, full-time job off the farm to be able to have the farm and then not being able to be a farmer, which is usual for most young farmers coming into agriculture."

Melissa Calhoun's sloping five acres look like anything but a farm. Her site description sounds pretty unappealing to the average row crop farmer—"limestone clay slope, north facing, scrub forest on half, surrounded by creeks"—but that's why it appeals to her. She's developed less than an acre for food and medicine, and she manages the rest for native perennial flowers for insects, birds, and butterflies, as well as medicine plants for humans. The five acres feel much bigger, having been divided from sixty-five when the previous owner sold. The owner of the other sixty acres manages it for hunting, and he is amenable to her foraging forays. She can hike and do her wildcrafting, and her neighbor is glad that she keeps an eye on his land when he's gone.

World traveler and freelance writer Lisa Munniksma is not interested in owning land soon. She says, "I really enjoying being able to support the local food system and work with farmers who already have something started, helping them improve and make sustainable food more accessible." She predicts that someday she'll have her own farm, but more likely for livestock and herbs than for a vegetable CSA. She thinks about it, and says, "Owning is very scary. Not being able to go work on a farm in Puerto Rico for the winter makes me sad. At the same time, having a piece of land where I could plant trees, strawberries, and be there in a few years to harvest, that's a really intriguing idea." If you have a gypsy spirit, it pays to make good friends with landowners. Lisa has a number of friends who have talked about farming together in the near future, and she will keep this option a possibility.

BODY

In a best-case scenario, you've got the land, the financial support, and now comes the real question: can your body do the work you want it to do? If that's tough to answer, you are probably a human being in modern society and you have adapted to your environment. If staying in touch with your body and monitoring its capacity to perform physical labor for

KNOW YOUR LIMITS

Women farmers, especially those working with animals, with men, or in profit-driven settings, sometimes push themselves harder than they should. The mindset of having to prove herself can cause a woman to overextend her physical capacity and can be counterproductive. The ideal is to bring sustainability into the mainstream, especially when it applies to how we treat our bodies.

Angela Wartes-Kahl and her team at Common Treasury Farm in Oregon were happy to receive a grant from the Natural Resources Conservation Service (NRCS) to buy and install a high tunnel (type of greenhouse). The timing wasn't ideal, however. The money came in halfway through the year, so they bought the greenhouse in September and had to build and install it by December. The seasonal help was leaving, so Angela had just a few people to pull off a job that really needed a large crew. Angela explains what they considered before jumping into it. "Being in our forties and close to fifty, we have to be careful about our longevity. We have to make sure that we're still going to be able to farm in the future. I don't want to face farming without an arm or a foot because we were pushing ourselves too hard."

Her farm's decision was to ask for an extension for another year, and she's glad that they had the time to do it right. "It just made it easier on ourselves. We could have pushed but probably would have hurt ourselves." Angela confides, "I see a lot of burnout, for sure, and I see a lot of people getting old way too early." She cares about herself and about the community, and she hopes the apprentices she mentors see the moral of this story. "People need to make conscientious decisions about how they are going to be farmers and what they are going to take on."

long hours is not required of you, then that is a skill that needs more development. This section provides some basic tools to help you get in touch with your body.

Physical Ability

One approach to farming is to view its activities and products as a health and fitness program. A great benefit to spending time in a garden or working outdoors is the improvement it makes in your own body as you work up a sweat, breathe hard, build muscles, and stretch your limbs and your limits—not to mention all the fresh food that fuels you. Yet, if we overdo it, we can experience a setback, such as a sprained ankle or torn ligament. We could end up less mobile for weeks or dealing with unnecessary pain.

Take some time to assess your body and encourage, perhaps even require, your partner, family members, or anyone working with you on your property to do the same. Think of it as an accident-prevention plan or, more proactively, as holistic and affordable health care for farm workers (and that means you!).

PHYSICAL ASSESSMENT #1
IN YOUR BODY

Adapted from the Nia Technique created by Debbie and Carlos Rosas.

Assess your body using the following guide. Do this on a yearly, seasonal, or monthly basis to check in with yourself and make adjustments for maintaining good health.

Date:______________________________Age:______________________________

Height:____________Weight:______________Resting pulse rate:________________

Medications:____

Health concerns:____

Under doctor's care for following conditions:______________________________

Stimulate self-awareness of your needs and capacity by answering the following questions.
Have you experienced any physical or emotional trauma in the past year?

Are you a care provider for someone else, such as young children or the elderly?

What is your stress level on a daily basis? What are your main sources of stress?

Do you experience chronic pain or immobility for which you are seeking healing?

What are your self-care practices for mental and physical wellness?

Daily:____

Weekly:____

Monthly:____

Seasonally:____

Annually:____

Rate each area of your body according to what you perceive in relation to each of the following sensations. Try a full range of motions to get a sense of each of these aspects: flexibility, stability, strength, discomfort, and pleasure. We are more likely to notice what is uncomfortable, so pay particular attention to what is pleasurable in your body as you concentrate on each area. Use a rating system of 1 to 5, where 1 = slight, 3 = moderate, and 5 = full or acute.

	Flexibility	Stability	Strength	Discomfort	Pleasure	Notes
Feet						
Ankles						
Lower legs						
Knees						
Upper legs						
Hips						
Abdomen						
Pelvis						
Lower back						
Middle back						
Upper back						
Rib cage						
Chest						
Neck						
Shoulders						
Upper arms						
Elbows						
Lower arms						
Wrists						
Hands						
Fingers						
Head						
Eyes						
TOTALS						

Where in your body would you like to improve your physical capabilities?

How do you imagine yourself making these improvements?

PHYSICAL ASSESSMENT #2

ON YOUR FARM

Let's apply your body awareness to your farm work. If you love doing it all, but your body isn't loving you back, maybe it's time to delegate. On the other hand, if you are motivated to get in shape through farm calisthenics, let's make sure you get a good workout.

What motions, chores, or tasks come easily or does your body enjoy? What tasks do you feel strongest when performing?

What motions, chores, or tasks do you avoid because of discomfort or weakness?

What motions, chores, or tasks do you need assistance with, but can accomplish with a helping hand?

What motions, chores, or tasks do you need someone else to perform?

What motions, chores, or tasks are you hoping to work toward doing? What steps will you take to become able to do these?

Every human body is made with the same basic structures, yet each person's body will move, respond, correct, and express in its own unique way. Many movement, martial arts, and healing art forms focus on awareness of sensations and of the body's constant adjustments in seeking balance. We can easily ignore our bodies until they speak to us in the language of pain. We also have the capacity to notice pleasure in our bodies and give it our full attention. What a wonderful way to experience the world and connect with nature.

Taking good care of your body, mind, and spirit is more important than any of the tools in your shed. Your body is a precious piece of equipment. You may be able to pass on your favorite hedge trimmer or your best hoe to a niece or grandchild, but your body will expire. The wisdom of self-care can be passed on, however. Share what you learn about tending to the garden of yourself.

At this stage—the assessment step—all you need is body awareness. Later in the book, we'll go deeper into the topic of self-care, along with some helpful insights from women farmers and holistic health practitioners.

FINANCES

The mapping exercise in which you sketched out the various layers of your property should have helped you take stock of your wealth of resources, which is much more valuable than money. Remember that all resources flow; this is especially true for finances. If you fit the definition of *hobby farmer* in that you don't concern yourself with making money from what you grow, this section will still benefit you by helping you take stock of resources to support the hobby you love. Later in this section, we'll explore what it would be like to scale up your hobby to justifiably call it a small business.

Are you farming to stock up provisions for your family or do you want to make a profit?

Money is a difficult topic for many of us. Some of that difficulty comes from social appropriateness and uncertainty. Would it be too

boastful to admit we made enough to break even selling at a market? Do we really know how much we spend on gardening? For many, our finances hold emotional weight, and sharing the details with strangers makes us feel pretty vulnerable. Others truly don't have a clue about their cash flow and wouldn't know where to begin.

Is the thought of getting control of your finances daunting, and actually making a living by farming unimaginable? It's not surprising that there's a stigma attached to this line of work, which is publicly cursed in our culture as one of the lowest-paying jobs. Don't give up. Take heart in the fortitude you probably already display in other areas of life—strengths that can translate into fiscal fluency.

Listen to Aunt Judy: "A strength of women farmers is that women run their own households, which has taught us administrative skills. We can handle several things at once, so we see the overall view of how things fit together and work together, and that grows as we take on more things. Women are already doing it in their own [homes], we just have to adapt it to whatever we want to do on the farm, whether it's husbandry with animals, maintenance, planning, or daily chores. Sure, we need to focus at certain times, [but] seeing the bigger picture keeps it better overall. We don't actually multitask, even though we think we do. It has to do with perspective."

Often, the thought of farming as a business is coupled with deep debt, but it does not have to be that way, especially if you begin farming as a hobby. *Sustainability* remains the buzzword when you think of financial priorities for your farm, and it will carry over to enrich your relationships and your lifestyle.

Resources

Here are some expert resources for understanding your finances and working toward your goals.

United Way

If you are starting from scratch or just need to refresh your perspective on personal finance, start in your local community. The United Way has at least one free and useful service that could help you. They can assist you in doing your taxes, starting a business, buying a house, getting educated about personal finance, and connecting with other local experts. Many areas offer a direct line to free help by dialing 2-1-1, a program that is supported by the United Way.

Among the wealth of resources found at www.unitedway.org are online calculators that can quickly assess your financial state and make recommendations (although you must

FINANCIAL SNAPSHOT #1
MONTHLY FLOW

To complete the financial portion of your assessment, gather your records and fill out the following charts to compose a comprehensive picture of your current financial state. If you are not currently tracking these categories, make estimates for now. Before proceeding, decide whether you want to track farming expenses separately from your personal expenses.

Monthly Income
Type of income
Total monthly income

Monthly Expenses
Shelter (rent/mortgage, taxes, insurance)
Utilities (gas, water, electricity, propane, oil)
Transportation (car insurance, gas, maintenance, repairs, registration)
Groceries
Clothing (include laundromat/dry cleaner)
Telephone/cell phone
Internet
Household (repairs, maintenance, savings for appliances, hired help)
Insurance (life, health, farm, homeowners')
Debt payments (car/equipment loans, credit cards)
Children/pets (clothing, toys, school expenses, daycare/babysitting, pet food)
Recreation (dining out, personal items, entertainment)
Gifts (tithing, charities, special occasions)
Health (gym, supplements, practitioners)
Hobbies
Other
Total monthly expenses

Total income $ ________ – **Total expenses** $ ________ = **Net monthly flow** $ ___________

FINANCIAL SNAPSHOT #2
NET WORTH

Assets	
Bank accounts	
Investment accounts (market value)	
Vehicles (resale value)	
Home (market value)	
Life insurance (cash value)	
Retirement accounts	
Business interests	
Personal property	
Personal loans receivable	
Total assets	

Liabilities	
Charge cards	
Personal loans	
School loans	
Automobile loans	
Home mortgage, home equity loans	
Back taxes/other unpaid bills	
Real estate loans	
Other debt	
Total liabilities	

Total assets $ ________ – **Total liabilities** $ ________ = **Net worth** $ __________

NATURE BARTERS

Bartering is not just for humans. Plants and animals live in mutualistic relationships every day, trading shelter for food, pollination for nectar, nurseries for protection. It's unnatural that we humans have to be compensated according to some arbitrary scale of worthiness. What would our society be like if we could just give and receive more freely? No matter what service you can provide, you have something to offer. That simple recognition can be life changing for some people, and there's no way to put a price tag on a sense of purpose.

provide the numbers, so you have to be able to estimate your income, expenses, and debt). This is a simple and basic start, which we often overlook when we are dreaming big. First things first—can you take care of your own needs?

Dave Ramsey

Dave Ramsey runs a media business based on coaching people through their journeys to living debt-free. His book *Total Money Makeover* provides a map with clear steps for building up savings and making planned purchases that make sense in the long run. His ridiculously simple envelope system for budgeting brings the emotional power of handling cash into focus. Ramsey's website sums up this effective method of budgeting: "Grandma's way of handling money still works. Many people used cash envelopes in the past to control their monthly spending, but very few do in today's card-swiping culture."

You can order Dave's books, listen to his podcasts, and order his special envelopes (or just make your own, to save a few dollars) on his website, www.daveramsey.com. Also, you can tune into his radio program and create your own envelope-cash budget system as you hear how other real people are successfully achieving their debt-free dreams.

Barbara Stanny

Ironically, Barbara Stanny, the daughter of the "R" in H&R Block, was totally in the dark about financial responsibilities until she had to face the reality that her husband was a compulsive gambler. Then he left. Suddenly, she had to grow up and get a grip on her own finances in order for her family to survive. She shares her hard-won wisdom in several books and coaching services on her website, www.barbarastanny.com. Her practical advice is linked to spiritual openness in making values-based priorities and disciplined, yet small, action steps.

Her book *Sacred Success* provides space to dream big, face fears, build wealth, and live up to your potential. Barbara's approach incorporates the powerful impact women have on their world, and compassionately acknowledges the awkward stages women may be going through in navigating a new paradigm. Her style of financial management prioritizes personal values and dreams by, as she puts it, "blending the feminine qualities of receiving and reflecting, cooperation and caring, with the masculine qualities of action and strategizing, assertiveness and toughening up."

After understanding your hopes and getting a grip on the reality of your current cash flow, you may consider investing. Barbara sums up all the confusing possibilities succinctly: "There are only two ways to invest: own or loan. There are only five places to invest: stocks, bonds, real estate, cash, and commodities." Knowing where your money is and where it is going empowers you to make decisions that will affect the legacy you choose to leave for the next generations.

If you want your farming hobby to pay for itself and possibly bring in a little extra income, potentially scaling up to become your primary source of income, then you must think strategically. If planning and planting a garden, breeding and raising animals, working with nature's timing, and preparing for the unexpected are already part of your farming duties, this should come easily. For many, strategic planning is a stretch from the regular 9-to-5, ho-hum existence.

Judy's advice comes from experience.

Barbara summarizes in her book what strategic thinking involves:

- Figuring the costs of doing business
- Creating a plan to keep costs low and revenues growing
- Cutting losses when something isn't working
- Designing effective structures and systems
- Integrating data and new information
- Daily/monthly strategizing and yearly long-term planning

Doesn't that sound like sustainable farming? Coincidentally, some of the permaculture design principles identified by Peter Bane in his book sound very similar and include the following:

- Producing a yield
- Self-regulating and accepting feedback
- Making no waste
- Designing from patterns to details
- Integrating, not segregating
- Cultivating vision and responding to change

Explore markets and enterprises individually so that you can break down the previous strategies into smaller projects. Finances are part of the picture, but they aren't the whole picture. If you can take the larger perspective of seeing all of the benefits of farming, and balance it with all of the costs, you can take that same skill and apply it to budgets and bookkeeping.

To begin tracking your income and expenses, you can tap into one of the aforementioned resources to obtain a budget template, or you can search for one you like on the Internet. Many software programs, such as Microsoft Excel or Google Sheets (which is free online), provide simple budget templates that you can customize to fit your categories.

Viewing your cash flow on a monthly basis is somewhat like going to the doctor and getting your blood drawn and tested. The results reflect your relative health, which will fluctuate daily but is generally either stable or unstable in some sense. Likewise, the flow of income and expenses reflects the condition of the assets and liabilities that you have accumulated but can mask chronic problems. Like making a change to your body's system, there's no magic pill that will bring everything back into financial balance. Celebrate the challenges you've overcome. Face any areas that you've been ignoring and accept that they are all part of your life experience. View your finances holistically, and see what they reveal about your values, priorities, time, and resilience.

Wealth

Wealth is not purely financial. Living a comfortable lifestyle means different things to different people, and sharing the surplus can take many forms. Graham Burnett writes about permaculture topics, and because this design system and philosophy is centered on

EXAMPLES OF CAPITAL

When I began a year of service through AmeriCorps, I attended a multiple-day training on the realities of poverty so that I could understand the ethics of service and apply them to my job, which was to support the development of urban community gardens. The training was held at a comfy four-star hotel, and they fed us three delicious buffets each day. We initiates were baffled at the expenses that the agency spent in order to teach us about poor people. Well, I was poor, and I stayed poor on the stipend they provided.

Looking back at that training, I realized the importance of a particular group exercise in which we explored the inherent meanings of poverty. We mapped assets for ourselves and the populations we would be working with. At the end of it, I did not feel poor at all, and I refer back to this exercise whenever I find myself feeling limited or stuck, especially when cash doesn't flow as much as I would like.

creating permanence in cultural systems, money has to be addressed sooner or later. In an article in *Permaculture Magazine*, Burnett shares some ideas to help integrate the cycles of earning and spending into land care and says that money is simply a representation of energy. "And like water, energy can flow about, cycling within a sustainable system, performing a multitude of functions and keeping things alive and vibrant. Or also like water, it can be stored and accumulated for its own sake, stagnating, stinking and building up dangerous levels of static, festering and becoming a source of obsession like a blocked bowel, damaging all that it contaminates." Money is a tool that must be used.

Burnett reminds us that there is more to life than money, and diversity in the toolbox is beneficial to the system. "In other words, a money-only based economy is a monoculture. And like a monocultural food production system, collapse is but one disaster away.... But a polycultural economy, like an edible landscape that has many elements, such as a diversity of fruit and nut trees, grains, vegetables, and so on, will be able to survive if one element fails."

Do Barter Systems Really Work?

Farming for money can be difficult, and we know there's much more to farming than money. Bartering can be a way to maintain the homesteading lifestyle while getting what you need to live comfortably. Bartering fits hand-in-hand with going back to the land. It means going back to a natural system that humans used for thousands of years before mass production of printed money. If you own cattle, you own the oldest form of money. Around 9,000–6,000 BC, when humans shifted from hunting and gathering to an agrarian lifestyle, livestock, grains, and vegetables became standards of exchange in various cultures.

When I worked as a park guide at Alibates Flint Quarries National Monument in Texas, I would teach visitors about the rock outcrops of flint, or agatized dolomite, a beautiful blue-red-tan variegated rock that could cut sharper than a razor. Fist-sized chunks of flint, called "blanks," were traded along migration routes stretching across the continent. These blanks could be knapped, or shaped, into any type of tool that people needed for acquiring food, shelter, and clothing. Whatever the end use of the flint, it had to be in the hands of people who knew how to use it, or it would be worthless. Those hands were most likely trained by elders who had learned, through living close to the land, to use what the land provided while keeping the land healthy for generations to come.

This brings me to my point: bartering is more than exchanging this for that. Bartering builds community spirit, fosters connections, and supports an interactive society. Bartering emphasizes identifying and matching gifts with needs rather than working for unidentifiable motives only to be rewarded with paper tokens.

Bartering is old school. Exchanging resources or services for mutual advantage means cutting out the middleman and judging value for yourself. Is bartering a thing of ancient history, or is it a system that can still work today? Is bartering only for idealists, the poor, or for those living

DID YOU KNOW?

There's a movement within permaculture, called *social permaculture*, that focuses on supporting the human systems, and some wise women are leading the way, such as Delia Carroll and Rachel Kaplan, co-founders of 13 Moon CoLab (www.13mooncollaborative.com).

off the grid or in radical communes? Can bartering become mainstream? Following are some ways to consider whether bartering fits into your life.

It's the informal, neighborly thing to do. Remember old-fashioned barn raisings? All the friends, family, and neighbors would gather to raise the roof on a new homestead. Although it seems rare these days, people do still need each other. Trading skills can be a great way to spend time with friends. Some of my friends in town have wood-burning stoves, and they take time helping split and stack wood at each others' houses. Some of them call it a "work trade," and others just enjoy it, have time to do it, and don't ask for anything in return.

It can be a one-off exchange. Bartering is a low-commitment agreement made directly with the other party. For example, I bartered with an animal acupuncturist: an hour of acupuncture for my cats in exchange for my taking photographs for the doctor to use in her marketing materials. Volunteering in a community garden one Saturday morning could mean that you walk away with training in a new skill and a few veggies, seeds, or transplants. A local Facebook page in my area, Kentucky Artisan Barter, connects artisans so that they can swap their goods and services with each other. Folks are trading things like rustic furniture for handmade soaps.

It's an organized system. A farmer can arrange community-supported agriculture (CSA) work shares to relieve herself of some labor by requiring each CSA member to work a certain number of hours each week or season in exchange for a certain amount of food. Teaching homesteading or gardening courses to CSA participants is another way to trade education for service. Permaculture training centers and independent growers offer experiences, such as building with natural materials, designing your landscape for sustainable agriculture, and foraging for wild foods. They will often charge a fee for participation but are usually open to full or partial work trades.

Organizations such as Time Banks (www.timebanks.org) manage barter systems and encourage a cultural shift around meeting needs in your community. Rather than trading directly with one other person, you bank your volunteer labor time and can cash it in by calling on the other members' services. One of Time Banks' core values is recognizing that much work goes undervalued or underpaid, so all time is seen as equal, and every hour of work is valued equally.

It can be a lifestyle. If you are ready to minimize your bills, leave home, and trade your time for food and/or lodging, you could find a farm on which to live and work. Worldwide Opportunities on Organic Farms (WWOOF), which links volunteers with organic farms, is the first organization that many young farmers try. Whether the arrangement lasts a week or a year, the standard agreement is full-time work in exchange for room and board.

More serious agriculture students may use ATTRA (https://attra.ncat.org) to locate a host farm (many of which are also listed in WWOOF). The National Center for Appropriate Technology (NCAT) maintains this database of internships, which lists expectations and contact information for farms that host interns and offer education as the main commodity of their work-exchange arrangements.

Help Exchange (www.helpx.net) is another option for trading work and living with others. This website opens the doors of large farms as well as bed and breakfasts, sailboats, hobby farms, and other establishments that need helpful tenants for a short time.

A caveat: When you start comparing your barter arrangement to hourly wages or weighing the value of one professional's time against another's, it can be a bit tricky. For example, would doing someone's laundry for one hour really equate to one hour of a professional psychotherapy session? In cases with a large disparity, both parties need to discuss what they perceive as fair and come to an agreement. It can't be emphasized enough that both parties need to clearly understand and clearly communicate the arrangement's expectations.

Additionally, if you plan to enter a bartering lifestyle, think about worst-case scenarios that could leave you homeless and without any savings to re-enter the cash economy. Don't let these fears hold you back from trading what you can, but be aware that you may need a backup plan.

ASSETS

What if you are short on land, time, money, and physical capabilities, but you still love to grow food, fiber, or medicine and share it with your community? You know you've got something to offer. Thank goodness the world does not revolve around money. Is that a shocking statement? Stop listening to market reports and tune into a different kind of capital.

Permaculturists love to take stock of assets. They view just about everything—even intangibles, such as spirituality and education—as resources. One of the principles that draws me to permaculture is that nothing is wasted, not even a random odd job. Such tangents can diversify the web of help you can call on when you need it. Plus, widening the scope of the assets available to you can really boost your optimism and fill you with hope for the future.

Ethan Roland and Gregory Landua explain what they see as eight different forms of capital that create something resembling an ecosystem. *Capital* refers to wealth, which can be money or any other valuable resource. If you are cash poor, you can still be very wealthy. Here the forms of capital identified by Ethan and Gregory with which we are all blessed in varying degrees:

- Intellectual: education, knowledge
- Spiritual: karma, enlightenment, forgiveness
- Social: influence, connections, favors
- Material: nonliving raw materials, buildings, tools, infrastructure
- Financial: money, investments, securities, precious metals
- Living: plants, animals, land, water
- Cultural: community, art, place-based connections
- Experiential: projects, real-life experiences, apprenticeships

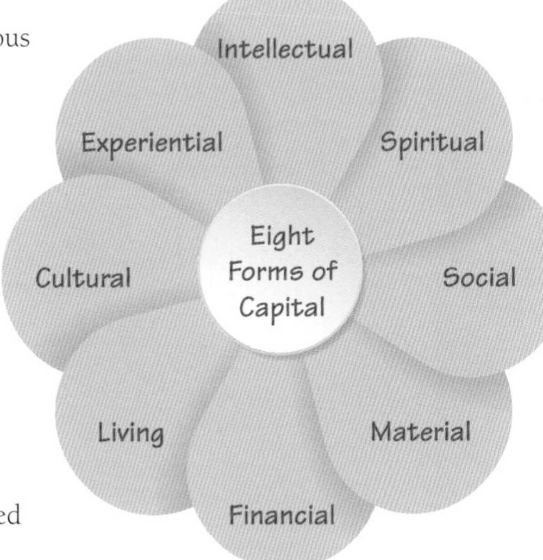

Take another look at the list and brainstorm. Write down who or what you can count on in each of these areas. You can complete this exercise for yourself as an individual, or you can map out your farm's capital, including all of the people involved in it and what they have to offer.

Who is always there for you emotionally?

Who are your dependents?

Who helps you make ends meet financially?

Repeat this brainstorm exercise, this time widening your vision to include those outside of your closest relationships. Who are you serving regularly in your church?

Who offers you fresh veggies out of their garden?

Who provides you with free loads of mulch?

Let's take a look at social capital in more detail to complete this overall assessment of your farm and its potential. By *social capital*, I specifically mean the people who form the primary relationships in your life. Begin with some more brainstorming. Give yourself time to write down the main relationships in your life and the types of support that you give these people and they give you.

Go back to your answers for each brainstorming exercise and assign a number to each person that you listed based on the following criteria:

- Those who are now directly involved with or impacted by your farming: 1
- Those who will be directly involved or impacted by your farming: 2
- Those who you wish would be directly involved or impacted by your farming: 3

Once you've given a number to every name, add more detail as well as any names you may not have thought of during your initial brainstorming. For each person to whom you assigned a 1, jot down their level of involvement and their roles (financer, moral support, recipient, consumer). For those to whom you assigned a 2 or a 3, estimate how soon they will become involved and what their roles will be. For all of the names, think about how long they will stay involved and how long their impact will last (years, lifetime, generations).

Your internal resources are only complemented by the external ones. It doesn't matter how much help is available to you if you don't ask for it. Look around you with eyes wide open the next time you feel a lack of something you need to get a job done. As Susana Lein said, "We are not islands."

A passion will only carry you so far, and then you'll need some expertise. For Aunt Judy, as a young newlywed with a baby, her love of her firstborn son and his aversion to anything but goat's milk propelled her into the role of goat breeder. She and her husband, James, dove into the farming world deep and built their lives on it. They worked at a boys' ranch, which was a residential community open to at-risk children in rural Texas. "At first," Judy says, "we did everything together—the garden and the livestock. We learned together, and we soon learned that some people and resources we had were trustworthy and some weren't."

For them, that learning came through experience and trial and error. After they lost their first goat and got run over by their first buck, Judy says, "We went to get a goat and found somebody who was wise." She describes her mentor as an "older, sweet fella" who became her go-to guy if she was ever in trouble. He taught Judy and James how to dehorn, how to manage a herd for the best production, and about breeding. Judy recalls their working relationship of give and receive. "He would let us borrow his bucks, which were really expensive, and we would trade out a kid, and he'd sell the kid. I got everything I could possibly glean from him, and that was a huge thing."

Maybe your best asset is the partner in your life who can complement and balance your strengths. Aunt Judy remembers growing up spending summers with her Aunt Mabel and Uncle Lloyd. Mabel didn't do the farming, but she was the administrator. Judy tells how the typical day would begin: "At breakfast, she would say, 'What are you planning to do today?' just in normal conversation, and he would tell her what he planned on doing, and she would say, 'And what are you doing with this?' So she would help him organize his day according to what was the most important thing to do—the priority—and he didn't even know she was doing it. She was the administrator. He was carrying it all out, the physical part, and he was really good at that. But they were working together in the whole."

Judy also thinks back on her interactions with customers at the feed-supply store where she worked as a newcomer to West Virginia. "Due to the fact that many small farms are not completely sustainable without an outside income source, many women have had to step up and fill roles that men have traditionally carried. They say, 'Necessity is the father of invention.' I'm of the opinion that if you have a passion to live on a farm, together with your gifts and planning, you can create a way. It requires intentional living whether you're male or female."

Joscelyn Elliott is young, a fresh college graduate, and she grew up feeling empowered to raise cattle. However, she still recognizes the reality of entering this field as a woman and the effort it takes to catch up with the learning curve. She says, "Women have to work a little bit harder because certain skills don't come naturally." Those skills probably didn't come naturally for the boys, either, according to developmental biologists. Joscelyn speaks for her generation and those before hers by stating the obvious: girls need to have the same opportunities that boys do to learn.

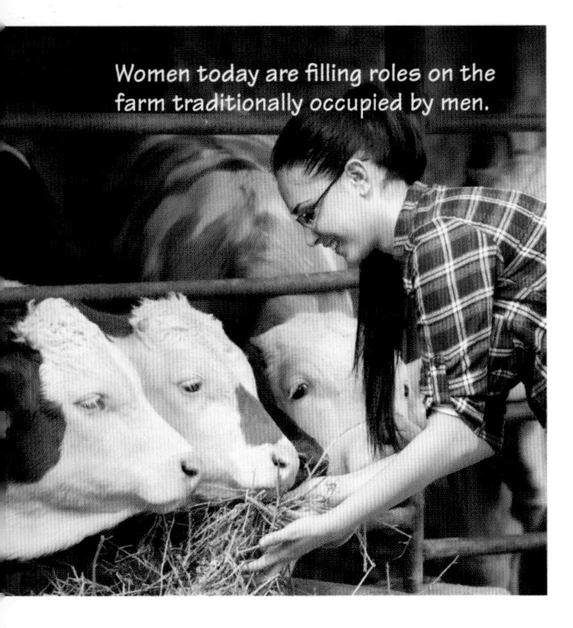
Women today are filling roles on the farm traditionally occupied by men.

In many areas, high-school shop classes are predominately attended by males, due mainly to the culture surrounding building trades. Joscelyn didn't let that stop her, and she took the class anyway. She shared an unfortunate but amusing

story about building a toolbox for her dad for Father's Day: "All the boys made nice ones, and the girls' were all uneven. My dad liked it anyway."

Kaitlyn Elliott went to the same school and shared the same memory. She laughed as she said, "My grandpa still has it displayed!" Displayed, but not useful. There's the difference. I asked if the classes made anything like that for Mother's Day. Wouldn't Mom love a toolbox made by her daughter? The room went silent as they thought about that. Finally, Kaitlyn answered, "Maybe a flower vase? Yeah, in our horticulture class, we learned how to make flower arrangements for Mother's Day." These young women know what they are up against, and Joscelyn firmly believes she can handle it. She repeated, "If you're interested in it, you just have to work harder at it."

Jessica overcame her share of obstacles on the way to becoming comanager of GreenHouse17's farming program.

That motivation is an asset that is difficult to quantify. Tapping into the gifts of our abundant consumer-driven society, we can find a zillion ways to repurpose materials, and the same can be true of misplaced energy. Redirect it toward your goals.

Jessica Ballard takes responsibility for finding her way through a weedy and winding path, with its obstacles and challenges. "If you are willing to do this work, and you feel called to it, you need to be open to what's out there." Jessica has realized and surpassed her goal to become an organic farmer after obtaining her degree in sustainable agriculture. She identified herself as "a single mom, no land, no husband, no anything." She remained open to possibilities, and she soon found herself farming in a variety of ways. "It was pretty sweet to land jobs at community gardens or doing garden education, and get compensated for doing that work."

Now she holds a salaried position as farm comanager at GreenHouse17, a shelter for victims of intimate partner abuse. She didn't expect the additional duties of tending to a dire social need while cultivating food and flowers, but that's what happened. She has held the attitude of "I'll do the work, and the next steps will happen." She explains further that "the universe put me right where I needed to be. It's pretty sweet to have a salaried job, but it's not just farming—there's all this other advocacy, trauma-informed care, and variables on top of the variables of farming. It's totally intense and stressful and not for just any farmer. But it is part of the service I give the world."

Chapter 2
Taking Care of Your Farm

So far, we have been narrowing down the wide world of possibilities to find out what you want to work toward and what you have to work with. We'll continue weeding out distractions so that you can focus on growing what you intend to and tending to what you grow. I invite you to use this chapter's space to

- prioritize what you want on your farm,
- begin or improve your designs,
- understand the capacity and limitations of various farming enterprises, and
- find the right tools to safely do the right jobs.

PRIORITIES AND BASIC DESIGN

We'll begin with another map, similar to and in conjunction with your site mapping and resource flows in Chapter 1. To do this, you will superimpose concentric circles on an overall sketch of your landscape, moving outward. These are your areas. They reflect the activity in, and the impact you directly have on, these spaces.

Use the following guidelines when deciding how large to draw your circles. Typically, the center circle, Area 1, will include your home base, where you spend most of your time. The next area, Area 2, could be the landscape you see directly near your home that you pass through multiple times per day. Beyond this, Area 3, would be your yard, garden, or crops farther away that you might visit often, but not necessarily every day. Area 4 could be field crops or your neighborhood, relative to how far your property spreads. If your property is large enough to have a Area 5 and beyond, these areas will ripple outward, representing spaces in which you spend the least amount of time and on which you make less direct impact, such as wild land that you may visit occasionally for personal renewal.

Sketching your areas is an exercise in attention to time. Where you spend your time reflects your priorities, regardless of what you may have written down earlier. Seeing this visual representation of where we put our energy can force us to get real about some of our daydreams.

Transfer a simple sketch of your base layer from Chapter 1 to the space on the opposite page. Beginning with your home, freehand-draw a circle that encompasses where you spend the majority of your time (when not away at work elsewhere). Label it Area 1. Continue with circles that cover the entire space and correspond to decreasing time and energy, perhaps allowing Area 5 to run off the page into areas beyond your borders. The circles don't have to be perfectly round, and try not to overthink this preliminary exercise. Feel free to make adjustments later.

Your Farm's Areas

Left: Area 1 is the area that you have the most contact with and most impact on.
Right: Your outermost area may even extend past your property's boundaries.

Do you see any patterns that surprise you? Are you really spending your time and energy where you thought? Does this change your priorities on where to invest time and energy on your farm?

Rachael Dupree is evaluating her new property, fifty acres of mostly wooded land. She says, "I wish I had that wisdom of 'this can wait until later' and 'this we should do now.'" She says that she is trying to use the principle of areas from permaculture: "starting around you and working [outward]." "But," she says, " I want to enjoy that wildflower pasture out there. So we need to do things so that we can get away and get out there."

Rachael is comfortable in the kitchen and is a budding herbalist, so she grapples with getting down to the far corner of the land, where a garden used to be. "I want our house to look nice, but it's more important to me that we start growing our own food, and that's farther away from our house." For her, taking it one step at a time is something she is learning slowly, with an open mind and resilient attitude. "We want to do everything right now. And we won't know until we start. So we're going to mess up a lot, and I'm OK with that."

GET SITE-SPECIFIC

Delia Scott worked as a horticultural extension agent and assisted inexperienced gardeners and farmers on a daily basis. They came to her for advice on just about everything, always worrying about doing something wrong and wanting to do it right. The cure for these worries, she says matter-of-factly, is that "you just have to go out and do it. What works for one person is not going to work for another. Everyone has their little unique ways of doing it, and that's when networking and talking to people helps—sharing that information. But it's just something you have to figure out for yourself,

really and truly. Building soil and when to start crops, when to start seeds, when to transplant—that stuff has got some leeway in it for everyone."

Aunt Judy worked at a feed-supply store, and she loved learning from the customers and the sales representatives. She remembered a particular producer of pasture seed who claimed that his product was the best. Her customers agreed. They had really positive results, so she was curious about why.

"I asked [the sales rep] what makes his particular product better," said Judy, "and he said that when he came to the area and did his rounds, he talked to the farmers and tested their soil. He determined the best mix for this particular environment and customized it specifically for [the farmer's] location." This is the value of working with local products, local human beings, and paying close attention to your own site's unique qualities.

As an extension agent, Delia helped countless farmers in her local area.

SOIL

Starting from the ground up is the only sustainable way to grow, whether it's a business or broccoli. The world's human population is multiplying at an alarming, exponential rate and consuming essential resources faster than producing them. What does this have to do with your hobby farm? If you look around the Internet for a few minutes, you'll find all sorts of claims that you can grow what you need and make an income on an acre or less. That may seem extreme, but one thing that successful intensive farming stories have in common is excellent soil health.

Soil Specifics

Physical: Understanding your soil begins with its physical characteristics. Texture, structure, and pore space make up *tilth*. Seasoned farmers know their tilth by its feel. Beginners can start with assessing soil texture by feeling for silt, sand, or clay. Follow the steps on the flowchart on page 75, and consult with a local soil expert to confirm your conclusions. Also, refer back to your layers map from Chapter 1, and find your county's soil map on the US Geological Survey (USGS) website (www.usgs.gov).

SOIL MATTERS

Why care about the soil? Because dirt is dead. Soil is alive! Soil is the most biodiverse part of any ecosystem. Millions of organisms can inhabit a spoonful of rich, healthy soil. Every arthropod, bacteria, fungus, or worm plays a role that affects the other members of the soil community. They shred, graze, parasitize, and predate on each other, but they mainly take care of organic matter. The underground food web processes everything from leafy tendrils to tough tree trunks. It also builds the infrastructure for plants' roots. It makes nutrients available, disperses water, and opens air pathways for good circulation, just as in any healthy community or habitat.

The benefits of biodiversity in your soil don't stop with the plants. Recent studies are finding that early exposure to healthy amounts of bacteria, fungus, and even some parasites could build children's immune systems, leading to fewer inflammatory conditions as adults. Scientists are trying to identify which members of a healthy gut microbiome affect specific problems, ranging from Crohn's disease to autism.

Similarly, researchers are often interested in isolating the cause and effect of certain bacteria and fungi on soil chemistry. However, according to the USDA Natural Resources Conservation Service (NRCS), "Many effects of soil organisms are a result of the interactions among organisms, rather than the actions of individual species. This implies that managing for a healthy food web is not primarily a matter of inoculating with key species, but of creating the right environmental conditions to support a diverse community of species."

Where do you find the richest, most diverse, and most resilient soil systems? In forests. Forests can have up to forty miles of fungus in just one teaspoon of soil, compared to several yards of fungus in a teaspoon of typical agricultural soil. Chemical-free gardening with native plants encourages a rich, biodiverse community above and below the ground and mimics the conditions found in the wild.

THE NRCS'S SOIL TEXTURAL TRIANGLE

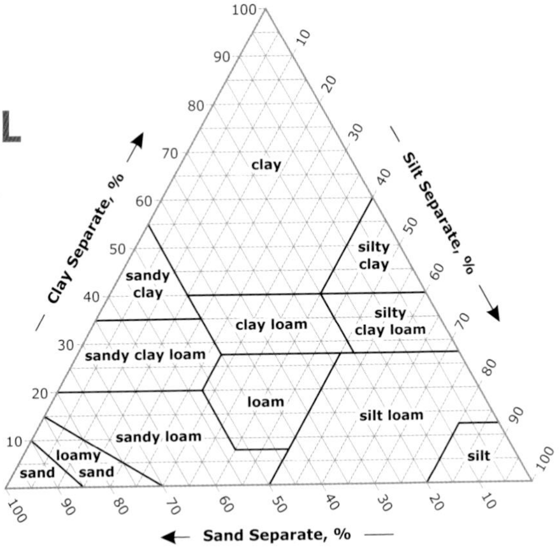

Image courtesy of USDA Natural Resources Conservation Service.

THE NRCS'S FLOWCHART FOR DETERMINING SOIL TEXTURE

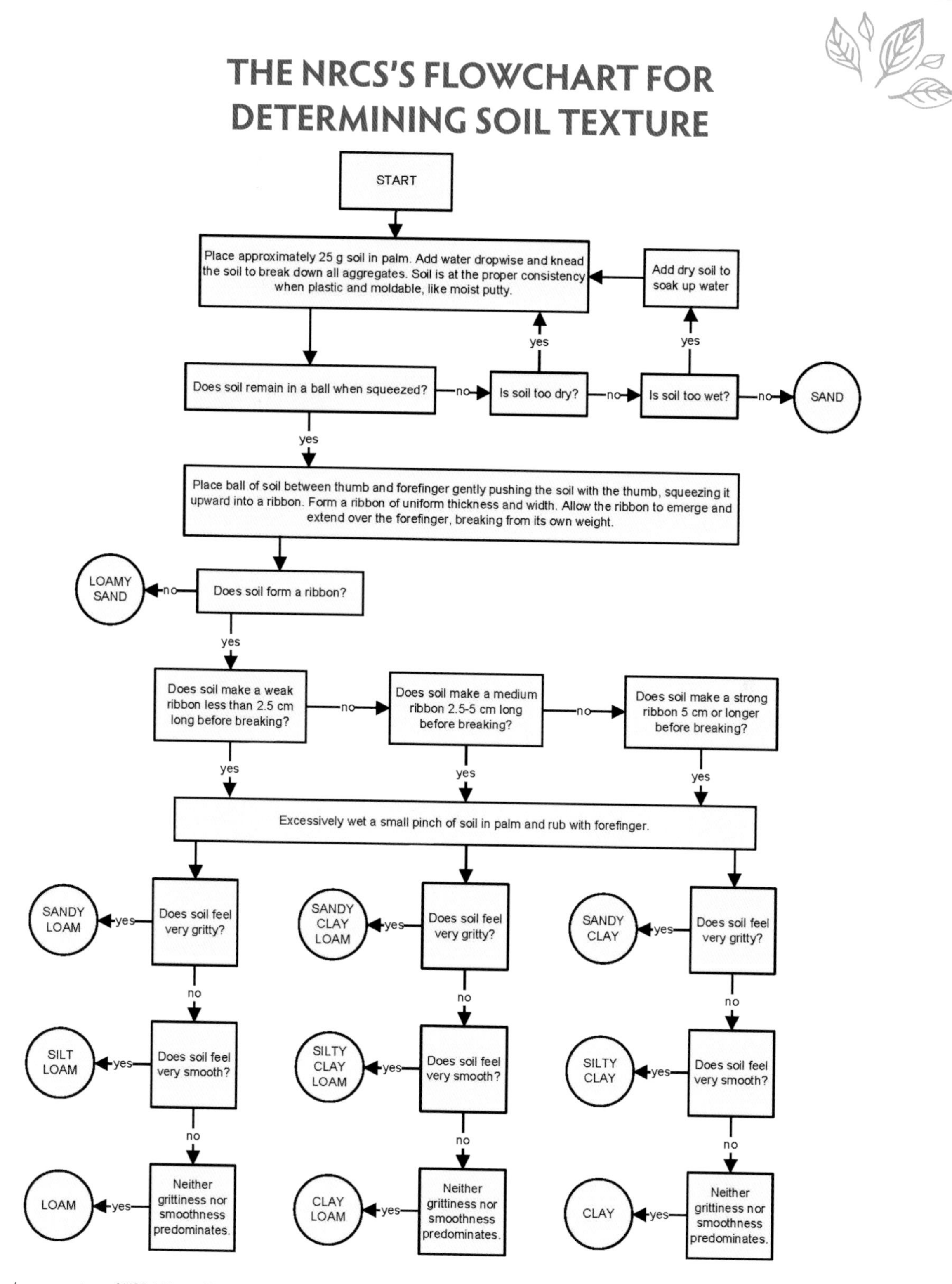

Image courtesy of USDA Natural Resources Conservation Service.

Chemical: As you are feeling around in the soil, get some samples that your extension office can test. Your local extension can provide bags and instructions for collecting samples. If you have an idea what you want to grow on the sites from where you took the samples, the local experts can make recommendations regarding soil amendments and their proportions. If you suspect that your soil could have contamination from heavy metals, especially in an urban space, ask your agent to connect you with a service to test for heavy metals.

Do your own research along with your extension agents' recommendations. The tricky thing about gardening and the mineral content of soil is that each mineral has its own personality and unique way of connecting with and becoming available for your plants. Too much or too little of certain minerals will help or harm different plants. If you want to build soil for its longest possible sustainable health, think about enabling it to feed itself rather than pumping it up with supplements when it is going through a weak spell. Biodynamic practices cultivate deep, healthy systems within the soil that minimize the need for amendments over time.

Biological: The beautiful thing about soil science is that the physical and chemical components are just dirt without the life that dances throughout the tilth. Microbes eat what falls off the roots of plants, colonize the roots, and live there. In turn, these microscopic bacteria and fungi help the plants grow in remarkable ways. What biodynamic farmer Jeff Poppen refers to as *microbial intelligence* enables the bacteria and fungi to get unavailable minerals and access them for the plant.

Slightly larger organisms, such as protozoa and nematodes, move nutrients around in the soil by eating the bacteria and fungi. Insects and earthworms eat those protozoa and nematodes and move the nutrients even farther. An entire community is working within the soil to provide what green life needs to thrive. On his blog (www.barefootfarmer.com/blog), Jeff says that these good microbes are "put out of work" by fertilizers. Encourage the proliferation of good microbes through practicing no-till gardening, which avoids compacting the soil.

The more you can interact with your land from a bug's-eye view, the better your foundation for growing will be.

Cultivation

The best piece of gardening advice I've heard, which applies to daily life as well, was this: always tend to what you have planted before you plant some more. Cultivating

GROWING ORGANIC

Angela Wartes-Kahl not only farms organically, she also coaches farmers through the transition to becoming eligible for organic certification. She said, "It's a lot more than just not spraying pesticides." Avoiding pesticides has, in itself, a benefit of magnitude because, for example, pests are the prey species of something else, so removing them upsets the system. Angela describes it as follows: "This new twenty-first-century agriculture is very holistic. Instead of the dominating aspect—for example, 'this weed is a problem, so let's get rid of it'— it's making decisions based on how something is going to improve my organic ground, soil, tilth.... How is it going to affect my whole-farm planning?"

The big picture is going to save our lives, to put it bluntly. Angela says, "In organic agriculture, it's all about the holistic view. And that is what's going to feed people because the soil is not going to die...organic ag is going to be able to produce food for millennia because it's treated well, and it's fed, and it's preserved."

Angela says, "Organic is always looking at the root of a problem, going back several steps, [and] looking back at what was done before." She gives an example of finding invasive weeds among her squash. She looked back at her crop rotations and thought either that she had put too many animals on the plot or that the weedy grass seed was brought in from somewhere else. Instead of finding out if she could spray the weed, getting the right herbicide, and figuring out how to use it, Angela would rather spend her time and energy understanding the conditions on her farm.

She adds, "Some farmers are figuring this out because they just can't afford it anymore. They are looking at [holistic practices] based on financial terms, rather than because we're stewards of the land." She concedes, "It's two different approaches, trying to come to the same point."

Angela often gets out in the field with conventional growers, and she is also a cotton inspector. She says, "I'm glad I get to see conventional aspects of farming. I want to understand what the rest of the world is dealing with, what is happening to workers on the farms who are exposed to herbicides, and how that's affecting their health over the long run. I also walk away from that and say, 'Boy, I'm really glad I'm an organic farmer!' I can drink the water and breathe the air. It's fantastic! And it's totally different."

ANNUALS TO PERENNIALS

Remy Hendrych has been working on her urban food forest for several years as she deepens her relationship with nature and sets the tone for the intentional community where she lives. "It doesn't make any difference to me, personally," Remy said in response to my question about planting annuals or perennials. She understands, as she says, "how annual vegetables have such a strong association with certain things—comfort, food, independence—and mean certain things that working with perennials does not mean in most people's brains—freedom, enlightenment, joy in living."

Imagine asking someone what she wants in the community garden, and she answers, "Chestnuts, groundnuts, persimmons, a huge array of wild edible greens, roots you can eat, Jerusalem artichokes, and medicinals." One might meet that answer with an uncomfortable blank stare. It's an unexpected mix when most people are looking for the same foods they can get in a grocery store—tomatoes, onions, lettuce, celery—so expanding our modern diets to include some of our ancestors' foods might be something to work up to. "All these things can support a way of life," Remy says. As a health coach, she acknowledges that a person's perspective is the key. "But if you look and don't see a way of life that is self-sufficient, it's not as attractive."

While she is happy to interweave kale and many other annuals amid her thriving fruit trees and native plants, she is enjoying reducing her workload as the perennials become well established. "I'm seeing that annual food vegetables are an important step in the path, in the arc of the story." What's the story? She says it's a path from A to B, and defines those points: "A is feeling totally disempowered and disconnected. B is more empowered and connected, wondering, 'Now, what am I going to do with that?'" It's a powerful question to lead her community into—a space that was identified as a food desert now has become an oasis, a food forest.

Similarly, Jessica Ballard is not preaching the gospel of permaculture outright, but she creates a living example that subtly inspires and provides a much-needed refuge for victims of domestic abuse. At the shelter where she farms, she admits that there's not much hype about the fruit and nut trees growing in the front yard. "A lot of people come in, and the last things on their minds are trees and forests." However, the idea of restoration agriculture resonates with her mission to nurture life—not to just keep it going in a sustainable cycle but to improve degraded conditions and leave the world a better place than she found it.

Jessica received a grant to reforest a portion of the forty-acre farm, and she planted a trial spot on a quarter acre. The next year, she applied and received the grant again, so the forest grew. A practical draw to restoration agriculture is that the plants selected are hardy and resilient for the region. "They should be able to thrive without me babying them," Jessica says. "They are doing well; we just mow around them and really don't do anything to manage [the area]. I based the design on Mark Shepard's model of polyculture guilds, which is based on plant families that live together in nature, taking the cultivated varieties that folks are drawn to." The hazelnuts, peaches, plums, blackberries, and oaks provide a quiet, shady space for residents, mostly women and children, to walk quietly or rest.

crops is about choosing your battles. Whether it's pulling weeds, working with volunteers, or salvaging or sacrificing the remains of an infestation, check in with yourself and understand what is worth the effort and what is not.

Aunt Judy reminds us to observe nature at work and try to get in sync. "If anything, living on a farm and interacting with nature is teaching you how things operate. If you're not in sync with the way it's operating, you're coming up against it. You're liable not to win too many of those battles. You may for the short-term, but it doesn't necessarily mean you will for the long-term."

Stay conscious of whether you are going about something because you've seen others do it that way and it's part of the culture you know, or if it's actually the best way. Judy says her husband, James, learned that, too. "Ranchers he worked with would insist he do it their way, but when he had the opportunity to do it himself, a lot of times he'd think through it and find a better way." Giving anyone an opportunity to apply critical thinking can overcome ingrained reactions to fixing a problem. Judy generalizes about the situations she's been in: "When men get in a really big hurry, invariably they do it push-and-shove." Well, what do women do? Judy laughs and replies, "When women get in a hurry, they put the kids somewhere and do it themselves or take the kids with them and do it themselves." No doubt she speaks from personal experience.

Aunt Judy reminds hobby farmers to get "in sync" with the way nature operates.

Cultivation is about daily maintenance and persevering through what can become monotony—checking in, taking a pulse, making adjustments, removing obstacles, making repairs. Technically speaking, cultivation is about loosening the earth and destroying weeds. Its goal is to ease growth for our intended plants. Aunt Judy's insight is again needed here, where we meddling humans love to overdo it. We are here to help regulate and balance nature's tendencies. Judy warns, "When we try to control too much of it, it becomes unmanageable, and we destroy it. There is a harmony in everything, and it's already been set up. A lot of times, we upset the harmony."

No-till farming advocates, such as Susana Lein and Jeff Poppin, would agree. Jeff would till only when working with compacted soil to loosen it enough to integrate

Jessica describes the "feminine vibe" of her work at GreenHouse17.

organic matter into it. He advises, "When you're tilling your soil, act like you're turning over a sleeping baby. Be gentle [and] go slow." Many common weeds sprout only in either disturbed or compacted soil, so try to make sure that you make use of the freshly tilled topsoil right away by sowing your seeds. Nature abhors a vacuum and will fill it with opportunistic pioneers.

Touching the earth connects us and affects us on deeper levels than merely the physical. The actions of yanking up a root, gently patting in a seed, and uncurling a leaf to reveal a camouflaged insect are acts that machinery separates us from. Perhaps this is why monocropping is perceived as a masculine, yang, forceful approach. Diverse polycultures that are designed on a human, rather than a mechanical, scale are considered feminine, yin, or receptive styles of farming.

Jessica Ballard notices these differences on a daily basis, tending flowers and vegetables with survivors of domestic abuse. "There's this inherent sense of sacred," which she describes as "the feminine vibe." Nearby, large equipment rumbles in and out of a construction site. She compares the bulldozers to tractors. "Oh, this is masculine—not male—agriculture. Machinery, the straight line, doesn't have to feel or emote; it's just a process of mechanics to go through and harvest."

Her work at the shelter employs intuition and cycles. Jessica says the healing space is enhanced by "the circle, the curves, the shapes of the earth, the elements, the important pieces that you are allowed to tune into." She can also be very grounded and pragmatic about the production aspect. "Your most tuned-in, feminine, magical, permaculture garden—you're not gonna get [anything] out of because you didn't control it a little." Jessica is not anti-tractor, but, she says, "the magic can't happen in those straight, rigid, objectified rows of corn and soybeans. You lose the whole sanctity of it. A little cultivation—yin and yang—is important to balance it."

WHAT'S YOUR FARM MADE OF?

By some definitions, a farm comprises both plants and animals. A hundred years ago, it was normal that every farm would have animals and plants both in production. Do you want or have animals, plants, or both?

Alvina Maynard shares her ignorance about raising alpacas. "I chose animals over plants. You have to know how to grow grass if you are going to have animals that eat grass. So that's a part that I am still learning." She is also learning that raising fiber sustainably requires a deeper relationship with the land. "That process doesn't start with manufacturing—getting your hands on the fiber itself—it actually starts with soil. Because if you take care of the soil, you take care of the grass. If you take care of the grass, the alpacas actually have something to eat. Nobody told me when I bought alpacas that I was actually going to be a grass farmer."

She has been self-taught and has found some excellent partnerships to support her new endeavor. She is trying to transition her pasture to organic by avoiding spraying the pasture. A farm nearby provides her with certified organic orchard alfalfa. She also has ambitions of doing intensive rotational grazing with electric fencing to help the soil.

How about you? If you want to raise animals, could you grow their food as well?

Whether raising plants, animals, or both, you must take care of the land.

The feel of mud squishing between your toes (but wait—what about worms?)… The dirt under your nails (but shouldn't you wash your hands before you eat?)… The joy of plucking a ripe strawberry out of the garden and biting into its warm sweetness right then and there (aren't you going to rinse it first?)…

Our industrialized world has become squeaky clean—and chronically ill. Many children today are unable to go outside to play because they have inflammatory conditions, such as allergies, asthma, or eczema. Many of these ailments can be traced to a lack of good dirt in our own bodies. The so-called problem may actually turn out to be the solution.

In a March 1, 2015 *Scientific American* article, researchers explain that they "now suspect that society-wide shifts in our microbial communities have contributed to our seemingly hyperreactive immune systems," says Moises Velasquez-Manoff, author of *Epidemic of Absence: A New Way of Understanding Allergies and Autoimmune Diseases*. "Drivers of these changes might include antibiotics, sanitary practices that are aimed at limiting infectious disease but that also hinder the transmission of symbiotic microbes, and, of course, our high-sugar, high-fat modern diet."

Worldwide, studies based on children's lifestyles are proving that early exposure to a healthy microbiome, the community of bacteria living in your body, is a key factor to having a strong immune system later in life. The Canadian Healthy Infant Longitudinal Development (CHILD) Study of 3,500 children born after 2010 found that babies delivered by Caesarian section were lacking certain "good" bacteria. Also, breast-fed infants showed an advantage in the richness and diversity of microbes living in their systems.

A European study gathered samples of allergens from the homes of children attending a Waldorf farm school and compared them with samples from more urban home environments. As expected, more dust mites, animal dander, and mold showed up in the homes of the farm kids. A separate study in Austria found that farm children suffer from significantly fewer allergy attacks. We can draw conclusions that children's immune systems develop tolerance to allergens when the children are raised with the allergens on a day-to-day basis.

Household cleaning can wash away beneficial microbes. A study published in the online version of *Pediatrics* in February 2015 found that homes with dishwashers had more incidences of children with allergies than homes in which the dishes are hand-washed. To compound that hypothesis, those living in homes without dishwashers were also more likely to eat farm-fresh and fermented foods. More complex laboratory research turned up evidence that Crohn's disease, autism, and anxiety are also connected to the health of our internal ecosystems.

Any farmer will tell you that the food she grows is only as healthy as the soil it grows in. The more biodiversity, the better. A gram of healthy soil could contain billions of microorganisms—bacteria, fungi, protozoa, and nematodes. They build structure and perform essential ecological functions below ground.

Friendly ecosystems full of anti-inflammatory microbes protect our bodies. "Our resident microbes seem to control aspects of our immune function in a way that suggests they are farming us, too," says Velasquez-Manoff in *Scientific American*. They live everywhere—your gut, skin, hair, couch, dog—and can affect everything—your physical, mental, and emotional state. Psychobiotics, a potential new pharmaceutical field, is finding correlations between what's living in a person's gut and their "gut" reactions to different stimuli. This could mean that having fun in the dirt as a kid could actually lead to a healthy system as an adult. However, untangling exactly which bacteria affect particular conditions will keep scientists busy for many decades.

Crowdsourcing is an innovative approach to connecting the microscopic dots. The American Gut Project, run almost entirely by volunteers, has accepted thousands of donations (financial and, ahem, fecal). For $99 and a sample, the project provides a personalized scientific report and analysis of your very own gut bacteria and includes your information in the massive database they are creating as this citizen science effort strives to understand our microbial and behavioral patterns on a population scale. As cofounder Rob Knight explains in a March 1, 2015 *Scientific American* article, "We have the potential not just to read our microbiome and look at predispositions but to change it for the better."

Just as the soil microbiome varies from field to field, so does the human microbiome vary from person to person. The Centers for Disease Control and Prevention (CDC) report that autoimmune diseases affect three times as many women as men. Some theorize that it could it be due to little girls staying cleaner and not being encouraged to get as messy as little boys. Subtle yet distinct boundaries between cultures also show up under the microscope. Compared to more primitive societies, such as the Hadza of Tanzania, industrialized Western diets and lifestyles reveal a reduction in variety and abundance of good gut bacteria.

Tony Stallins, associate professor of geography at the University of Kentucky, has been trying to translate the biology and sociology research into geographic language, and he draws some healthy conclusions: "People think they are going to be able to finalize this relationship between us, our environment, and our bacteria. But things will always evolve. The medical world depends on this stability, so patenting this may not be possible."

Remy Hendrych is a health coach and nature-based mentor who lives with Crohn's disease. She ferments her own foods and draws on traditional wisdom to guide her diet and lifestyle. "Personally, it's one of the few foods I think I can draw 'conclusions' about in my own health—that with fermented food, there seems to be a strong correlation with a number of improved health markers. I have also seen this in other people I've worked with who have been sick and are healing."

Do you need to improve your own microbiome? You've got some options. You could take probiotics. Look forward to the new dirt movement, when pharmaceutical companies commodify your microbiome and sell you beneficial bacteria supplements to target your specific condition. You could make and eat fermented foods, such as yogurt, kefir, and sauerkraut. You could pay close attention to the quality of the soil from which your food comes. You could invite your friends out to the garden or field with you to plant some potatoes in the rich tilth or harvest some spring onions.

Professor Stallins muses about the cures for the ills we've created: "We can simulate dirty fingernails, with all the possible side effects, or we can just go out and get our fingernails dirty." The side effects of that might just be pure, childlike joy.

Plants

Because animals depend on plants, and plants depend on soil, we'll explore some possibilities and key concepts when looking at plant options. In a resilient system, everything serves more than one purpose. When it comes to harvesting food from plants, plants are also serving as soil stabilizers with their root systems and as air purifiers that convert carbon dioxide into oxygen. These benefits come with just letting nature do its thing. The same benefits can be created intentionally by understanding the ecosystem advantages that different plants provide.

Some aspects of gardening stem from a multidimensional and multifunctional approach that includes using perennials when possible, rotating annual crops conscientiously, using mulch wisely, composting, cover cropping, companion planting, and seed saving. Permaculturists try to use as many of these techniques as possible in one space, which is called *stacking functions*.

The idea of stacking takes shape naturally in a forest, with layers of vegetation that protect, support, and benefit each other. The tallest trees create a canopy that filters sunlight for the shade-tolerant understory, which grows just above shrubs and woody plants, followed by wildflowers, herbs, ferns, and down to ground cover, such as moss. Vines rise up through all of these complex layers and penetrate from the ground to the

Beans, corn, and squash—the "three sisters" of the garden.

canopy. Of course, the vertical aspects of the forest also descend underground, where the deepest depths of native grasses' taproots could surpass the heights of some trees.

Legumes, such as peas, offer benefits for everything growing in the garden.

To bring the same idea into the garden, a simple comparison is the "three sisters" garden. Traditional wisdom in early agrarian cultures taught that combining beans, corn, and squash—the three sisters—provided balanced nutritional components. Corn lacks two essential amino acids that beans contain, and the seeds of squash contain good fats that the other two lack. Additionally, the corn provides a structural support, a natural trellis for the beans to climb. The squash spreads its wide leaves out below as a ground cover that cools and protects the soil, acting as a type of living mulch.

The way that legumes transform atmospheric nitrogen exemplifies another mutually beneficial system conceived and perfected in nature. The seemingly magical process can be attributed to a symbiotic relationship with bacteria present in soil called *rhizobia*. Nodules on the roots of legumes provide a habitat (shelter and food source) for the bacteria, which convert nitrogen into forms that the plants can actually use. Not only does the leguminous host benefit, but the surrounding plants can then draw the bioavailable nitrogen from the soil as well. In a sense, the legumes are freeing up resources that all in the plant community need to survive and thrive.

Non-legumes planted in rotation also continue to benefit. Besides fixing nitrogen, legumes feed on soil microbes and leave behind beneficial residues. The Food and Agricultural Organization (FAO) of the United Nations reports that cereal grains and oilseeds that follow one or two seasons of *pulses* (edible seeds of legumes) can reduce their carbon footprint by up to 34 percent. Improved soil health means less fertilizer and lower costs to the farmer.

Animals

Animals are a part of your farm, whether you call them livestock or not. The soil is alive with very small animals that recycle and transport nutrients. The air is a pathway for animals that visit flowers and disperse seeds. The ground and vegetation provide habitat for many critters that do their best to keep other populations in check. *Wildlife in Your Garden*, a book I cowrote and edited, describes the ways in which we host a variety of wild creatures in our yards.

Intentionally bringing animals into your care requires dedication and an understanding of life cycles, which goes beyond the simple provision of food, water, shelter, and space, and that is a fine starting point. Just as you have evaluated what resources you have for growing food, repeat the evaluation exercise with a mind toward the animals you envision or already have on your farm. Do you have what it takes to raise them?

With all creatures, great and small, the fundamental idea is to begin with the end in mind, which is one of Stephen Covey's *7 Habits of Highly Effective People*. This applies to animals directly because every action you take to house and feed them is for an end purpose. It could be food, fiber, manure, lawn-mowing services, companionship, education, or a mix. Hopefully, it will be a mix, because animals are like any other element of a thriving system, and they serve more than one function.

What animals do you currently have?

List your animals' purposes, their functions, and how they benefit the whole system.

What are their needs?

What animals would you like to have?

What purposes would these animals serve? What are their functions?
How can they benefit the whole system?

What are their needs?

BALANCING FARM AND ECOLOGICAL CONCERNS

An inspiring conversation with one of my seventh-grade students went something like this:

"Ms. Lanier! I know what I want my science project to be!"

"Oh, really?"

"I'm going to create a riparian buffer and restore the stream ecology on my grandmother's farm. She raises cattle on about 500 acres, and they are polluting the stream and eroding the bank."

"Oh, really?"

"Yeah, and I'm going to create a floating wetland, too, so our pond will filter out the excess nutrients."

"OK. Wow. Let me know how I can help."

How many of you reading this even know what *riparian* means (along a stream bank) or know what a floating wetland is (basically a human-made island that provides wildlife habitat and improves water quality)? Thanks to a land-based school curriculum, along with outreach programs from universities, extension offices, and environmental education organizations, thirteen-year-olds like this one can speak the language of restorative ecology. He applied for grants and calculated the cost of moving fences farther from streams and stabilizing the soil by planting a food forest. He came to me daily to share his progress and ask for more literature to take home. His grandmother learned about impervious surfaces and stormwater runoff. Generational gaps are narrowing, and ancestral wisdom of the land is collaborating with scientific understanding on how to manage farms for long-term ecological stability. This gives me hope for the future.

Thinking about my own education at Instituto Terra regarding farmland restoration, I witnessed and documented the difference that one reforestation project in Brazil is making. A decade of replanting the Atlantic tropical forest on what had become a degraded and barren ranch has stimulated the economy and refreshed the environment. Streams that had dried up now flow again, wildlife has returned, tourism has increased, and environmental education programs attract schoolchildren and train new conservation technicians in field ecology.

Considering the balance of income and ecological restoration, profits can support the work but do not necessarily make us better stewards of the land. Most of the truly land-conscious farmers I know have spent time in third-world countries, where they learned how to live within their means. They returned from the Peace Corps, mission trips, studying abroad, or WWOOFing with a new perspective. Now overwhelmed and perplexed when facing the vast aisles in the average American supermarket, they have

one thought that rises to the surface: what a waste of resources.

The new generation of agroecologists refuse to accrue debt and buy massive machinery to grow commodities. They are downsizing the industrial ideas of farming, and they are choosing to grow subsistence crops in urban farms and community gardens. They welcome families into their CSAs, share tools, and host potlucks, which is not so different than the way their great-grandparents did things.

Back at school, my seventh-graders discussed an article entitled "A Village that Planted Its Rain and Watershed" from the book *Rainwater Harvesting for Drylands and Beyond* by Brad Lancaster. In a Rajasthan desert, Laxman Singh and his organization, Gram Vikas Navyak Mandal Laporiya (www.gvnml.org), pushed the idea of shaping the land to maximize rainwater harvesting. He designed earthen tanks that worked with the landscape to slow the sparse flow of rain and let it hydrate their crops, trees, and animals.

The community eventually rallied to support Laxman's methods and enforced rules for cutting down native trees. The punishment fits the crime: plant another tree, write an apology, and pay eleven pounds of grain. The semiarid village of Laporiya has sprung to life. It is a model for permaculture and land stewardship, not because of more rain but because of more efficient use of the land and the water.

Farms are places to restore ecology. When systems flourish, they provide a surplus. When needs are met, energy is made available. The question is not about striking a balance. If we carefully tend to our resources, the scales will be tipped in favor of all life on Earth.

The sheep paddock on Angela Wartes-Kahl's Common Treasury Farm.

Keeping the end in mind also refers, quite literally, to the fact that you will likely outlive your livestock. Dr. Temple Grandin has turned the modern processing facilities upside down with her unique perspective on animals' end days. Even factory farms and their associated slaughterhouses can show humanity and make those final hours as peaceful as possible. Dr. Grandin has her Ph.D. in animal science, but her real knack for understanding them comes from her autism. She can relate her own sensitivities to the ways that animals perceive. In her book *Animals in Translation*, she writes, "A great deal of my success in working with animals comes from the simple fact that I see all kinds of connections between their behavior and certain autistic behaviors." She works with producers and processors to at least improve, and ideally understand, the conditions where cattle and hogs are kept.

As of 2006, one-third of all American cattle and hogs were handled in facilities designed by Dr. Grandin. Her method is to identify the root of a problem and address it. She explains, "The principle behind my designs is to use the animals' natural behavior patterns to encourage them to move willingly through the system. If an animal balks and refuses to walk through an alley, one needs to find out why it is scared and refuses to move. Unfortunately, people often try to correct these problems with force instead of by understanding the animal's behavior." Some of the considerations she emphasizes in training handlers include the facts that nervous mammals generally respond well to gentle pressure, that prey animals feel threatened by movement above them, and that poor lighting or reflections can trigger a fear response.

As part of having autism, Dr. Grandin had to learn how to empathize with other people in order to communicate and socialize effectively. She describes how she translates this to the animal world: "When I put myself in a cow's place, I really have to be that cow and not a person in a cow costume.... I place myself inside its body and imagine what it experiences.... I have to follow the cattle's rules of behavior. I also have to imagine what experiencing the world through the cow's sensory system is like."

Aunt Judy echoes this observation of patterns, which she applies to all of farming, in particular to working with horses. She considers communication techniques. "How do

you speak to a horse that doesn't speak human? You have to devise a language skill." She has observed that horse language often revolves around dominance and submission. There's a hierarchy that involves a great deal of trust. Judy explains that in a herd, there is one lead mare that dominates the others. "When I'm going out to train, I have to establish who I am.... They push on you, and they are dominant, so you learn why they are doing it. Or there will be a fear horse—something scares them, and they push in. Watch to find the lead mare and how she speaks to the herd."

Judy knows that learning how to communicate with the animals requires dedicated time and attention, and it's the reason why horses are not for just anyone. "So many people who are raising [horses] just as a hobby don't know that, so they get run over. They don't know what's going on with the horse."

Many of the women I interviewed have a lifelong love of horses and go to great lengths to have them in their lives. Many farmers keep animals around for other reasons than food or fiber production. What are the other services that animals provide?

For Helen Terry, her donkeys are important landscapers and night watchmen. She says, "I don't have to worry when I'm here alone at night. I don't have an alarm system. I know my donkeys will make noises if there is anyone on the property." While she loves them and provides a refuge for them, they do a job as well. "They do have practical purposes. They keep our grass down, and they help me feel protected."

They are also helping Helen's farm-based retreat business in a big way by just being cute on social media. Helen reports, "Any time we have a post with donkeys on it, it gets double the response of anything else we do. People seem to have a bond with donkeys." Even her guests who don't have any farm experience are drawn outdoors to the pastures.

HELPFUL RESOURCES FOR CONSERVATION PROGRAMS

USDA's Farm Service Agency
www.fsa.usda.gov
Click on Programs and Services and choose Conservation Programs from the drop-down menu.
USDA's Natural Resource Conservation Service
www.nrcs.usda.gov
Click on Programs.
Web Soil Survey
http://websoilsurvey.sc.egov.usda.gov

It wasn't long ago when Helen and her husband, Joe, were total beginners at working with large animals. "When we bought the property, it came with two horses that the previous owners didn't know what they were going to do with. As we were signing the paperwork to buy the property, [the sellers] casually said, 'Oh, we have these two horses that come with the property. Would you like them?' Joe and I were, at that point, naive and inexperienced with animals. The sellers said it would be very easy and didn't cost very much to maintain them. So we took on two horses without any experience. I now know better."

That was only the start. The previous owners also let four other horses stay on the property, and those horses' owners showed up one day with two donkeys they had rescued. Helen recalls, "They said, 'We think you should have donkeys on the ranch.' We were quite shocked!" Eventually, Helen and Joe got two more donkeys, and word got out around Montgomery, Texas, that these new greenhorns were rescuing donkeys. "Before we knew it, we had seven more donkeys and another horse."

What is Helen's definition of a rescued animal? "Not one that we went out and bought. People approached us, saying they'd identified that the animals were at risk and needed a home." The situation for some of these nonfood animals can be dire. When Helen and Joe bought the ranch in 2009, it was a desperate drought year. She recalls, "The cost of hay went from $3 to $18 in one summer. Farmers couldn't afford to feed their animals, and donkeys were the first ones to not be fed."

Helen sums up the benefits that her braying alarm system provides. "The big reason is what I call 'cosmic salary.' People share that they've never touched a farm animal before. People love getting the scraps of food from Joe and going out and feeding the donkeys. We've got more photographs of donkeys than anything else on our ranch."

Soma Ranch's "alarm system."

Empathy plays a big role in the production of both meat and wool from alpacas at Alvina Maynard's ranch. If she didn't care about other humans and animals, she would probably not be in the alpaca business at all.

"Have you heard of slow fashion, slow clothes?" Alvina starts her guests off on a tour of River Hill Ranch with this question. Just as the slow-food movement is a response to the fast-food industry, slow fashion asks consumers and manufacturers to scrutinize the sources

ARE MEN REALLY STRONGER THAN WOMEN?

When I asked women farmers about the qualities men bring to farming, they usually responded in terms of physical strength. However, it doesn't have to be that way. Katie Ratajczak is working on her Ph.D. in gender studies and says that some of our notions of men's strength are results of our cultural expectations. In the animal kingdom, the males of a species are usually smaller than the females because the females must carry and nourish the young. At some point in our history, humans began altering this tendency. Katie informed me, "In Anglo-European culture, boys were favored over girls. They were provided more resources, were pushed to do more, and nutritionally gained more resources to do more. In turn, women had less and became smaller." The culture pushed those size differences over time, and they had little to do with biology. Bodies adapted to do the work they were asked to do.

Machinery replacing manual labor and other health and behavior factors are affecting that discrepancy, leveling out the size differences. Katie says, "Now, in the United States, resources aren't as divided like that. The gap is closing." The tallest man in the world is not much taller than the tallest woman.

Like wearing high heels, the color pink, makeup, and wigs, all of which men used to do primarily, nothing stays the same in gender. It's a nice working environment when the differences take a backseat to working in community. Delia Scott, when thinking back on her twenty years of gardening professionally, laughs about how hard she tried to keep up with her male colleagues. "I wish I had known that it's not a competition!" She explains, "When I started working at a nursery, I felt all this pressure to perform as well as the guys. I pushed myself physically in ways that were not good." She advises others to take note that male culture may be more competitive, and women are likely to pick up on that. But, just remember, "You don't have to push yourself to the point of exhaustion."

Women I talked with agree that there are some manual tasks they might ask a man to do, but, in general, they can find a way if they are given a chance. On the farm where I worked part-time, we females would make a few more trips with lighter loads and help each other lift heavy objects. Often, men would walk up and take over, doing it for us, which caused an immediate shift in the camaraderie and teamwork atmosphere. It went from "Yeah, we got this" to "Hey, let me do that for you." A subtle message comes across that the males may lack confidence in our ability to find a sustainable approach to both achieving the goal and easing the strain on our bodies.

DO IT HERSELF

I have received two tool kits in my life. The first one was when I graduated high school. A family friend gave me a Do-It-Herself tool kit with the basics—hammer, pliers, wrenches, screwdrivers—and they all fit neatly into a portable, handy case. At that time, I had mixed feelings about receiving the present. On one hand, I felt acknowledged and a bit delayed, as if this were a rite of passage that I didn't really know about and almost missed out on. I was eighteen years old, and this was the first tool set anyone had given me, even though I had played with my brother's toy tool bench as a toddler, taken shop class in middle school, and assisted elders with car repairs on more weekends than I cared to. Thanks for the great present, but why hadn't I been trusted with my own hammer before?

On the other hand, I felt ashamed. The friends gathered for my graduation celebration kidded me about the Do-It-Herself brand. What would happen if I let boys use the tools? Are they made especially dummy-proof for girls? I took the ridicule in stride and joked along with them, but I always wondered if the female versions of the tools were just more cheaply made, similar to kids' gardening tools that break after a few uses.

Additionally, I felt confused. The silliest thing about this gift was that the toolbox itself was a shade of baby blue. Why wouldn't they go all-out and make the girly tool box pink? Why not put flowers all over it? It was as if by making the box blue, they had disguised the blocky words exclaiming that this was a female version of what males don't need a label for.

Ten years later, I worked as a tour guide at a national monument that preserves ancient quarry pits where stone tools were created. I learned what a tool really is: something that every animal carries with it in order to survive. Mountain lions are equipped with sharp teeth and claws. Ravens use their highly adaptable beaks to manipulate objects. Owls glide on wings made of sound buffers for a silent and stealthy flight. Bison use their hooves to turn the soil where the prairie seeds that cling to their fur will drop. Nature is full of tools. What did humans have for survival? Not a very furry coat, only average eyesight and hearing, pretty dull teeth, and a lot of other inadequacies. But we had been given a relatively large brain. So the tool we carry around with us is the tool that figures things out. Our brain.

Ironically, the second tool kit I received came from the superintendent of this national monument. He had been my boss through some major life changes: my marriage and then the death of my mother. When I left the park, he gave me a going-away gift. It was a Leatherman multi-tool in a beautiful leather case. My name was engraved on the tool. This is a hefty tool with a super-sharp blade, and I can do almost anything with it. It folds up neat and tidy and can ride on my belt. I kinda love it. I also kinda love my Do-It-Herself kit. The Leatherman stays with me or nearby most of the time, and the baby-blue kit is always in the car.

of their clothes. The supply chain on a T-shirt can impact everything from human rights policies in foreign countries to water quality in your own stream. Alvina makes this message clear when she hosts tours and sells at farmers' markets.

Her care for the animals extends to wanting to utilize them fully. As they age, their fiber quality degrades, and it's time, Alvina says, "to send them on to the great green pasture in the sky." She raises the dual-purpose species to provide a lean meat that is high in iron and protein. She can relate to the surprised reactions of some people. "A lot of people say, 'How can you eat them? They're so cute!' Well, you come out and do herd health with me, and you deal with the eight-year-old, 200-pound, not-nice alpacas, and it makes it a little easier. It's never easy…but at a certain point, they end up costing me more money than they are making me. They are livestock. This is a farm business, so they get sent on."

Alvina softens a bit after stating these facts, admitting that she gets very attached to the fluffy friends. Her two small children help out on the ranch and are closely involved with the animals and the alpaca wool processing. She explains, "The meat my family consumes—I know where it's from [and] how humanely it was treated. As much as it hurts, I stay with them 'til the end."

TOOLS: LINKING FARM AND SELF WITH THE RIGHT STUFF

Tools, in the broadest sense, are the bridge that links farm care to self care. When we have the right tools to do the right work, there's a flow that eases the stress on our bodies. In this section, we'll take a look at basic tools, including protective gear and items that make work easier.

What's in your toolbox? What are some of your must-have, favorite tools?

FARM-GIRL FASHION

After your brain, the next most important tool is your body, and protecting it from the elements will keep this tool useful. Think of the clothing choices you make for optimal functionality to keep this tool in proper shape. What you wear on a given day will depend on the weather, what risks you are facing, and the work you are doing.

FEET

Your feet are the hands that touch the earth, all the time. Unfortunately, there's no single magical footwear that protects, aerates, warms, cools, allows you to sense the earth under your feet, and costs very little.

Billie Jean Geers speaks on behalf of other animal caretakers and says that her favorite piece of gear is her muck boots. She goes out to feed cattle, pitches hay to goats and sheep, walks horses, refreshes pigs' water troughs, and collects eggs every day at the historic farm site where she has worked for ten years. She has both summer and winter muck boots, and she feels that they are worth the investment because they allow her to take care of the animals without being concerned about what she'll track in later.

The footwear you choose must be able to handle the terrain, the soil, and the weather.

Joscelyn Strange deals mostly with horses and cattle, and she loves her Ariat boots. When asked to describe them, she replied, "They're just like a cowgirl boot—definitely not tennis shoes!" Delia Scott works in many different garden and farm situations and says, "I used to shell out a lot of money for waterproof things. Nothing is waterproof after a certain point." She is fond of Blundstones, Australian pull-on leather boots, for their versatility, noting that they hold up really well and there are no laces to mess with. Other footwear I've seen on my women farmer friends ranges from finger-toe shoes to Chaco sandals to Keen closed-toe sandals to 1980s-style moon boots to hiking shoes to regular old tennis shoes (sorry, Joscelyn!).

While protecting your feet is most definitely a worthy investment, farmers can be rough on their gear. Shop at secondhand stores if you tend to go through shoes quickly.

Whether you are chopping weeds down with a slingblade or shearing sheep, attention to your body's needs is always important. For example, rocky soil may necessitate footwear with ankle support. Softer shoes might work better if you are frequently squatting and flexing your feet. Simply assess the situation and take care of those "hands touching the earth" with every step.

HANDS

Hands are extensions of our hearts, moving through the chest, shoulders, and arms to reach through our digits. Gloves can be a gardener's best friend, again depending on the work you're doing and the situation. Delia Scott says, "Gloves are kind of pointless after a certain point. I'll go out and weed something in the backyard if I just need a break from working. The feel of dirt on your hands—after five minutes, you just don't notice it, you get so engrossed in what you're doing."

Personally, I prefer gloves when I'm picking tomatoes, squash, okra, or anything that might make me itchy. Same goes for wearing long sleeves. Also, I dislike the smell of tomatoes on my hands. Maybe that's something you enjoy about gardening. I don't mind getting compost under my nails, but it is not my goal to toughen up my hands, so I generally wear gloves if feeling the texture is not important. However, when I'm potting,

Gardening gloves offer protection from sharp branches and thorns.

NATURE'S RULES IN GREENHOUSES

The golden rule of permaculture is simply to work *with* nature, not against it. Your greenhouse will be your studio as nature's creative partner, not as a dominating owner. Remember that natural laws still apply indoors. Wind direction and speed, water flow and drainage, soil quality and microbial diversity continue to rule the lives of the plants you choose to grow, so choose wisely.

Plan ahead when you select plants. Visualize and measure the space they will need when they reach their full, mature height. Many dwarf varieties of fruit trees produce abundantly without raising your roof. Create plant guilds, which are groupings of mutually beneficial plants. *Forest gardening* means stacking layers of plants that fill canopy, understory, shrub, herb, vine, and root niches, just as in a wild forest. Perennials will sustain the greatest yields over time, while annual vegetables and flowers have their place in the herbaceous layer.

Shane Smith's book, *The Bountiful Solar Greenhouse,* points to the importance of noticing the microclimates that your plants inevitably create and how you can turn any perceived problems into solutions, which is one of permaculture's basic tenets. Shane explains, "Here's a common example illustrating the different microenvironments. Think of a cool, relatively dark place in the greenhouse where there is only enough heat and light to produce plants halfway to maturity. Let's compare Swiss chard and tomatoes. Now, what do we have to eat with an immature tomato plant? Nothing. We cannot eat tomato leaves or blossoms. How about an immature Swiss chard plant half the size of a mature plant? There are still plenty of leaves to eat, although they are smaller. This is the difference between utilizing environments for either productivity or non-productivity, at least as far as our stomachs are concerned. So, it's a matter of plugging the right plant into one of the available environments at the right time."

Recreate your vacation paradise, but be careful with mixing and matching exotic locales. If you are creating a tropical greenhouse, don't include plants that are susceptible to fungal diseases. If you are creating a Mediterranean climate, understand that many varieties can withstand a freeze because they go dormant, but tropical plants cannot.

Also pay close attention to your water's quality, pH, and minerals. Your plants don't like to be splashed with cold water any more than you do, so use thermal mass water tanks to warm your irrigation water.

Building healthy soil is as important indoors as outdoors. Soil temperatures are more crucial than air temperatures, so monitor them closely. For minimal inputs, companion-plant nitrogen-fixing plants with nitrogen-loving plants. Remember to be careful with overwatering because roots need good ventilation, too.

Let animals come inside. Your greenhouse ventilation should allow pollinators to fly in and out. Incorporate worms to aerate your soil. Chickens, ducks, and rabbits can add rich fertilizer to your garden beds. To further enrich your plants' lives and your own, bring music and birdsong into the space. Permaculture greenhouses are not sterile laboratories. They reflect the spontaneous flow of life in the wild.

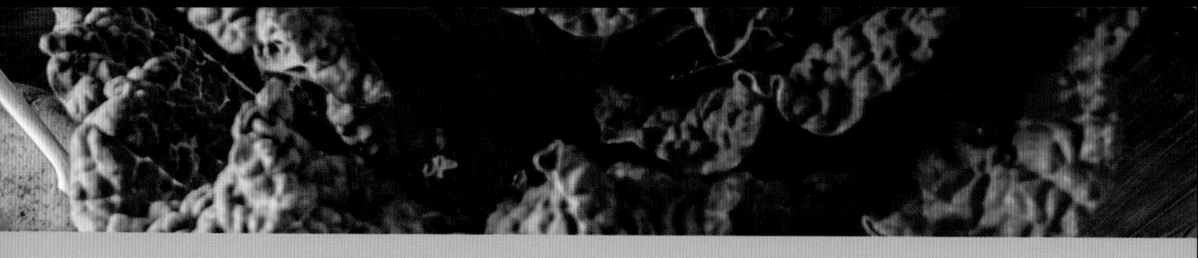

Don't be intimidated by thinking about all of these factors at play. Jerome Osentowski runs a nonprofit organization, sells edible landscape plants and tinctures, teaches workshops, and designs greenhouses for a living. For him, he says, "This is not a hobby. This is serious." But he encourages anyone to "just start with the simplest season extender: a cold frame, a high tunnel, or some kind of hybrid. Just start somewhere."

If you're ready to add an indoor oasis to your property, look for Jerome's book, *The Forest Garden Greenhouse* (Chelsea Green, 2016). He shares decades of experience with greenhouse design and how, as he says, "In the most difficult environment, we've created a paradise, with no major resources and very little money."

Ideally, a greenhouse is a functional place but also a fun place. The lines between the permaculture trinity of caring for the earth, caring for the people, and sharing the surplus are blurred, and you simply enjoy hours at work or play in your own Eden. Arriving at this merger does not happen overnight. It will be an ongoing process of making adjustments and learning as you go. Bringing nature indoors is about cultivating a relationship with the organisms that you share your space with. When you spend time day in and day out with living things, you notice the nuances that, over time, create a bond. Just as any relationship needs to grow, you will learn and adapt with the seasons.

sowing seeds, mixing soil, or working with herbs, the gloves come off. I need to feel the texture and sense the subtle moisture, fine hairs, and aromatic traces on my fingertips.

Just as machinery doing the labor for us has impeded our direct relationship with the earth, gloves can, too. In her research with southeastern US farmers transitioning out of tobacco production, Dr. Ann K. Ferrell writes in her book *Burley: Kentucky Tobacco in a New Century* her observations on the stigma around wearing gloves: "The loss of physical contact is representative of the changing relationship with the crop." Stripping tobacco—pulling the cured leaves off—used to be a hands-on activity that the farmer and his family took care of themselves.

On larger-acreage farms, stripping is only done by the hired help, who wear cotton gloves for protection. Dr. Ferrell writes, "The gloves are as important symbolically as they are tangibly. Gloves protect the wearer from the sticky residue of the cured tobacco and keep hands warm in cold stripping rooms. But tobacco men don't need protection from the crop or the cold. Gloves also block the wearer from full access to the plant, inhibit natural movement, and lead to rougher treatment of the leaves; gloves lessen the ability to handle the crop gently (with respect). Not only do gloves provide a literal and symbolic separation from the crop and therefore tangible evidence of the changed relationship with the crop, but they also symbolize the move to tobacco production as 'business,' as it is primarily hired hands that are protected by gloves."

Don't get me wrong. Gloves are incredibly helpful to prevent splinters, provide a barrier in case of stinging insects, and provide good insulation in colder temperatures. I often use gloves just for the extra grip they give when I'm using hand tools or carrying buckets. I could work on building up my calluses, or I could just slip on my trusty old gloves. It depends on the task at hand.

SUN PROTECTION

One of my favorite tools is my big straw hat. I found it for $12 at a thrift store—what a deal!—just before moving across the country to begin a farm apprenticeship. It's been my trusty companion through four summers and shows a little wear and tear, but it really makes a difference in keeping my eyes shaded. Delia adds her comments on sun care: "The importance of sunscreen. Nobody told me that. Always wear a hat. People who have been in agriculture for a long time always wear long-sleeved button-down shirts for a reason. It actually does keep you cooler."

It's true. In the hot summer sun, you'll create your own evaporative cooler if you wear long-sleeved, light-colored tops and long pants. I've seen professional landscapers soak their shirts in cold water before putting them on to go work out in blistering heat. Use what nature has given you—sweat—as a body temperature regulator. Keep in mind that if you are not sweating and you are working outdoors on a hot day, you are probably dehydrated. Drink plenty of water and take frequent breaks in the shade.

Consider a shade hat mandatory gear for working outdoors.

SIMPLE TOOL GUIDE

Gender equality and history aside, modern women in the United States are built differently than men. We usually have a lower center of gravity and more strength in our legs and hips than in our arms and torso. Tools, however, are mostly designed for an average-height person with good upper body strength. Don't buy cheap, flimsy tools, Delia warns. "Always use good ergonomic tools because [otherwise] you can injure yourself. Just use tools that you're comfortable with and suit your body type." Favorite garden tools mentioned by the women I talked with include the following:

- D-handle shovel, which provides leverage more easily than a shovel with a standard handle
- All-purpose soil knife (a.k.a. hori hori) with a serrated edge for digging, weeding, and planting
- Broad fork for small scale, no-till use where the farmer employs the "lasagna gardening" technique to build topsoil
- Slingblade with sharp edge on both sides for cutting tall grass with minimum effort
- Walk-behind tractors with multiple attachments (but only those that are necessary)

Your best tool could also be an expert to help you make a tool purchase, such as Melissa Calhoun. She is the shipping manager at Earth Tools, a small family business nestled in the Appalachian foothills that ships and receives tools from all over the world. Melissa described some of the most popular and useful tools she sells as well as how she directs customers to the proper tools. She starts by asking customers a few questions to help them determine their needs and narrow down their decision. There's a tool for every job. Her questions include:

- What kind of work are you doing?
- What kind of soil do you have?
- How often are you doing the task?
- How big and strong is the person who will be operating the tool?
- How much experience does the tool operator have with this type of tool?

Pruning shears are on a farmer's must-have list.

HAND TOOLS

One of the most popular hand tools among female farmers is the garden knife, or soil knife, or hori hori. It serves many functions, such as digging and slicing. Be very careful with this knife because the blade can be very sharp. You should keep it in its leather case when not in use.

Pruners are a close runner-up. Many varieties are on the market, and it's best if you can try handling them before purchasing because the grip can vary widely from one model to the next. Melissa recommends pruners that swivel, especially for florists or anyone doing a lot of pruning. Needle-nose shears are a light and easy-to-squeeze tool for reaching and snipping small stems. Go for a professional-grade pair of shears because good quality will make a big difference in reducing the stress caused by repetitive hand motions.

LONG-HANDLED TOOLS

While it might seem counterintuitive, shorter people can do quite well with long-handled tools, depending on the job. Hoes with long handles save your back from bending and pulling out small weeds, and the sharper the blade, the less force is

The stirrup hoe counts removing weeds and aerating soil among its uses.

required. The ergonomic design of some hoes includes a side handle for extra leverage when pulling the tool toward your body.

The standard square-head hoe is useful for wide spots, and a variety of lighter hoes with narrower heads are also available. A trapezoid hoe with extra sharp corners can gently lift out tiny weed seedlings from between tightly spaced baby vegetables. A stirrup hoe is sharp on both edges and swivels slightly to allow for efficient cutting and pulling as it uses energy going both away from and toward the body.

For digging out rocks or deep roots, use a D-handle shovel with a wide base for placing your foot. The shorter handle and D-shaped grip provide a lower angle for you to leverage your weight.

A broad fork will loosen up and aerate soil in the spring without tilling or turning. It can also gently upheave root crops when it's harvest time. I think the broad fork is one of the most fun garden tools to use because it allows you to keep your back straight as you rock back and forth on the wide base (like standing on a swing).

CUTTING TOOLS

It might not be a common sight among either urban farmers or growers of field crops, but a scythe can be very useful to cut knee-high or higher grass. If you have chickens, for example, you can cut the day's forage from your homegrown seeds. You can use it instead of a weed trimmer for selective weeding around the yard. Scythes come in various handle lengths and curves, and Melissa emphasizes the care needed when swinging the blade. A longer handle may be better, but be aware of your surroundings and how far out the blade is reaching as you work through a stand of tall grass.

In the realm of restoration agriculture, the hazelnut tree can certainly pull its weight. Imagine growing protein in your garden on a plant that could easily grow in the wild without any maintenance. Imagine growing protein on a plant that is so deeply rooted that it can survive a drought. Imagine growing protein that you only have to plant once and can harvest for twenty-five years.

Imagine harvesting that protein, grinding it up, and spreading it on a slice of toast, maybe with a little chocolate. Yum! The tasty ingredient in Nutella could likely be a gateway plant into restoration agriculture. The hazelnut.

Wild hazelnuts (a.k.a. filberts) grow throughout the Midwest, in the eastern United States, and in parts of California, Washington, and Oregon. There are two wild natives: the American hazelnut (*Corylus americana*) and the beaked hazelnut (*Corylus cornuta*). Many cultivars are commercially available.

A few years ago, I was involved with planting some food forests on urban tracts of land where blighted properties had been torn down. Our challenge was to provide foods that were recognizable, low maintenance, appealing to the neighbors, adapted to our climate, and nutritious. For community gardens, in which caretaking is inconsistent, perennials are ideal. Berries are a great fit, but quality proteins and fats are harder to find in a food desert and are so very essential. We decided to try planting some hazelnuts, and they have done well.

At the home or community-garden scale, volunteer gardener Ann McCarthy offers a bit of advice from trying hazelnuts for a few years. "One key is to make sure that the cultivars you get can cross-pollinate each other, so buy from a knowledgeable source. And remember, they are pollinated by the wind blowing the pollen from one to another, not by insects. So plant the different cultivars near each other and in a spot with a bit of breeze."

In a small way, these shrubs are restoring urban landscapes while they restore community and provide some essential needs. Behold the basic concept of restoration agriculture.

To a large extent, modern gardening mimics Big Ag instead of nature. Based on annual crops, the robotic and linear repetition of plant, harvest, replant not only creates a monoculture, it also ensures monotony. Agricultural systems based on annual crops have been taken to extremes, and the larger ecosystems have paid the price for machine-centric standardization and convenience in the fields.

In response, the concept of restoration agriculture harkens back to wilder roots. Finding inspiration from nature itself, this method uses perennials to restore soil health and ecosystem stability while providing an abundance of nutritious, real food in a beautiful forest, providing what our bodies and spirits hunger for. Not to mention the wildlife habitat they provide, too.

If you have trouble envisioning how a wilder landscape can work if humans are intended to participate in it, Mark Shepard's book *Restoration Agriculture: Real-World Permaculture for Farmers* describes scenarios that make it all seem very possible.

In his introductory chapter, Mark describes desertification of the Midwest, America's so-called breadbasket. He contrasts a wasteland of lifeless, depleted soil with a vibrant permaculture farm. He describes the sensations of life surrounding him, from the softness of the earth to the sounds of frogs and toads. He spots several types of birds and butterflies, and "a host of other insects chirped and trilled in

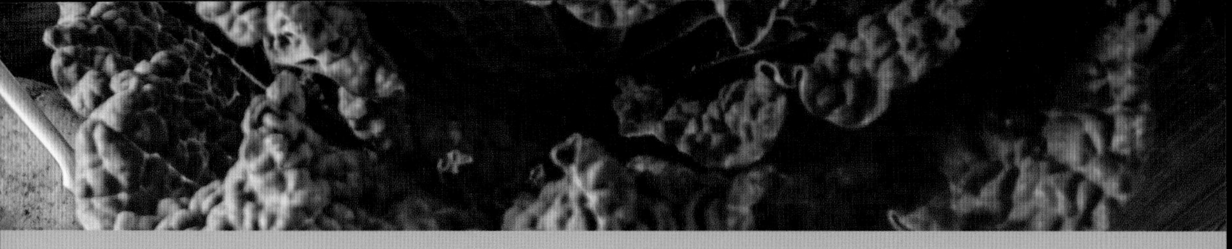

a delicious cacophony that made me smile." But it wasn't just a walk in the park, he writes. "In addition to all of this, I was surrounded by food! The hazelnut shrubs I passed had already been harvested, as had the cherries, mulberries, kiwis, and pears. Apples in a bounty of red and gold were being harvested into pallet bins while the nearby chestnuts finished ripening, awaiting their turn for harvest. Several steers grazed on the abundant grass and pigs snuffled about under the hazelnut bushes looking for dropped precious gems." The farm he describes had previously been "a bare-dirt cornfield," but it had been healed.

Mark goes into detail about different plants he recommends for this transformational process, and hazelnuts are near the top of the list. They are an overlooked yet abundant wild food with some very appealing qualities. They are an average-sized bush with average-looking leaves, but the nut kernels are very high in vitamin E. The oil can be pressed out of the kernel, and the remaining nutty meal is 30 percent protein. When processed sustainably, the nut hulls themselves can power a processing facility, producing practically no waste. Mark writes that shells burn hotter than wood and even coal, and much cleaner. Of interest particularly to urban gardens where heavy metals could be present in the soil is this factoid: "Hazelnut shell ash has been shown to bind and make unavailable several heavy metals, most notably cadmium."

Commercial production of hazelnuts is gaining steam as well, but the wild plants require a little taming to produce consistently and in more geographic regions, especially in response to global climate change. The Arbor Day Foundation landed a $3 million grant from the USDA to improve hybridization of hazelnuts, picking the best from the wild and domesticated varieties. With the help of a consortium of growers, the Arbor Day Foundation has been testing cultivars and selectively breeding for cold hardiness, drought tolerance, and resistance to a fungal disease called eastern filbert blight. The group's website states, "If successful, the expansion of hazelnut production through the Consortium's research will give the United States the potential of becoming the world's leading sustainable hazelnut producer and address critical issues in areas of agriculture, environment, wildlife habitat, society, health, hunger, and sustainable energy" (www.arborday.org/programs/hazelnuts).

Hazelnuts, like all woody plants, can also be a source of biomass, and they require some regular coppicing (cutting back to encourage new growth) . Trimming the branches mimics the natural browsing of animals and rids the plants of diseased areas. Compared to annual agriculture's routine of harvesting entire plants, prepping soil, rotating crops, and replanting each year, a little annual trimming seems like a minimal investment for the rich rewards of densely nutritious nuts. Mark optimistically reports that the practice of coppicing hazelnuts in Spain has kept the same grove producing nuts for 1,600 years.

Restoration agriculture aims to use abandoned or otherwise marginal land to add to the food supply without imposing agriculture on wildlands, a serious issue to consider as biofuel crops are replacing fossil fuels and land is required to grow them. Perennials of all sorts have been studied for their potential for biofuel production, and the US Department of Defense is very interested in hazelnuts because they produce the most oil per acre of any other perennial. Could hazelnut oil power the vehicles and heat the homes of the future?

A variation on the standard scythe is a serrated scythe, which helps grab and cut. A machete has a straighter blade and can also make a good weed trimmer. A billhook is like a machete with a slight curve that can be used for cutting high branches, with various reaches made possible by different handle lengths. Melissa doesn't sell her favorite tool, a slingblade; she was gifted hers by an elder. It works like a scythe but has a double edge.

WALK-BEHIND TRACTORS

A walk-behind tractor is a mini-tractor that keeps your feet on the earth. If you are envisioning a souped-up rototiller, you're mostly right. A walk-behind tractor has two wheels and is very versatile because it accepts a variety of attachments. At the risk of reinforcing gender stereotypes, I think of it like a food processor. I may want to chop, stir, dice, or blend, so I put on different attachments depending on the desired result.

On a tractor of any type, the power take-off (PTO) attaches the accessory to the motor. This is the area you have to be most cautious around and where most farm-equipment-related injuries occur. However, Melissa assures customers that walk-behind tractors are very safe to use. The machine is designed to work in your favor, with handlebars that can swing around and attach on the other side.

A walk-behind tractor can be fitted with a range of attachments for different types of work.

On Melissa's steeply sloping, rocky, and densely grown property, she has only had trouble with the tractor sliding downhill. She asserts that they are safe and easy to handle. "It's certainly safer than a chainsaw or weed trimmer. It's definitely user friendly. There are a lot of women out there who are farming with these things."

An example of a really handy accessory, out of about thirty possibilities, is a flail mower. It chips and grinds and is closer to the ground than a deck mower. Melissa explains, "Things like sorghum, corn, tree saplings, turn into dust."

A tiller is another attachment. Melissa says, "A common thing would be to get a rototiller and a grass cutter—what we call a sickle bar mower. You can do a lot of space in a short amount of time." She goes on to say about farmers and their tractors, "People get fancy with 'em. There's a chipper, shredder, snowblower, bulldozer—that's more like you've got a huge budget and you've already got a tractor that's doing work for you."

IN THE GREENHOUSE: THE SUN IS NATURE'S POWER TOOL

When you think about making the most of what nature gives you to work with, remember that some of the tools you need are provided by nature itself. Extending the growing season by capturing the sun's energy is a multifaceted solution to the problem of winter.

Imagine winter in the high Rocky Mountains, a blustery wind chill of 8 degrees whipping your parched skin. As you step into your tropical greenhouse, all of your senses shift dramatically. You remove your coat and fogged-up glasses inside this 80-degree-Fahrenheit, 80-percent-humidity oasis. You breathe deeply, drinking in oxygen-rich air, and your head feels clear. A dense, musky aroma—a mixture of moist soil, herbs, and fruit—soothes your dry nostrils. The flagstone patio under your feet invites you to unlace your winter boots, roll out a yoga mat, and take a long winter's stretch.

This is the scene that Jerome Osentowski has created in his tropical permaculture greenhouse, Phoenix. Phoenix is featured in books such as *The Permaculture Handbook* by Peter Bane and Toby Hemenway's *Gaia's Garden*. Permaculture practices on display throughout this thriving environment include stacking layers, forest gardening, maximizing edges, capturing and harvesting energy, producing no waste, and multiple functions performed by every feature in and around the building. Permaculture gardeners embrace Mother Nature's systems and attempt to replicate her genius and efficiency in their designs. "It's like creating an ark. We sail into the tropics, or the Mediterranean, while eating figs, and it's net zero in terms of energy," Jerome muses in "Glass Houses," an article in the January/February 2015 issue of *Hobby Farms* magazine.

Year-round growing is a hot topic. Minimizing the miles your food travels, eating fresh produce, and spending time around green, living things doesn't have to stop with the end of summer. A greenhouse is not only a microcosm of plant life but also a testing ground for permaculture principles and your own imagination. In a permaculture greenhouse, the ethics of care for the earth, care for the people, and share of the surplus overlap and support each other in every design element. You can kick back and relax here while feeling good that your self-indulgence is a self-sustaining system.

EXPLORING NATIVE ROOTS

Native plants and natural landscaping may seem like they don't fit in an urban space. They aren't typically associated with agricultural crops. Some are too big for small spaces. Some (like rare orchids) can be intimidatingly particular and only survive under specific conditions. Worst of all, if you replace your front lawn with pollinator habitat, what will the neighbors think? For me, living in the city and crossing these barriers to urban wildness has been a slow and clumsy journey, but so worth it.

My foray into gardening with native plants has been mainly through living vicariously through others. I like to hang out with native plant experts, listen to them, go on walks with them, photograph things they point out, and gradually let the information sink in. I am surprised at how much I've absorbed this way and how that knowledge will resurface at odd times. Sometimes, I will see a particular flower, and its name will pop into my head without my even really trying. Other times, I stare at a common leaf and can't summon the name. This is sort of what natural landscaping is like. Unpredictable.

I have made the common mistakes that all newbies to native plant gardening have made. First and foremost, by calling myself a gardener, it implies I have control over something growing. That's not so much the case with natural landscaping. After all, I do love the plants for their inherent wildness.

I can stand back in awe of the colossal spreading sunflower that consumed my entire flowerbed at my downtown apartment complex. I planted the unassuming little green sprig, excited to bring it home from a plant exchange, wondering why nobody else wanted this native sunflower. Had I simply flipped the tag around, I would have seen the operative word, *spreading*, and had a clue as to what its tendency

A monarch butterfly on spreading sunflowers.

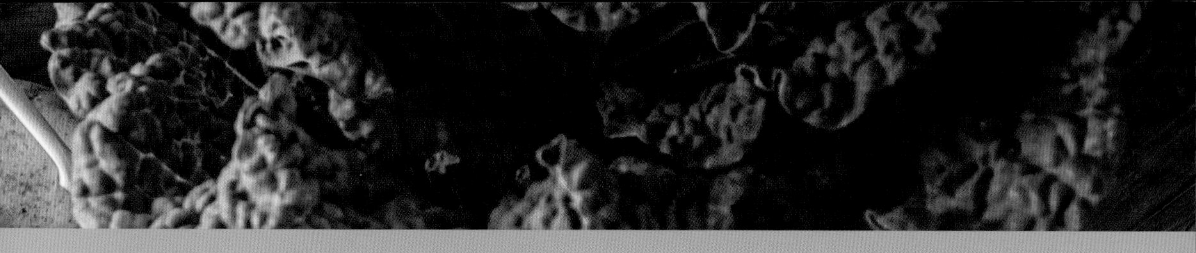

A volunteer moves wild ginger plants.

would be. So the eight-foot-tall stalks angle awkwardly toward the sidewalk and wait until just the last minute of the summer to finally bloom and prove to all the neighbors that they are in fact an intentional planting and not a weedy fire hazard. This pretty yellow flower attracted monarch butterflies, so I love it despite its assertiveness. It is a close relative to the sunchoke; it's extremely prolific, with little baby sunflower shoots coming up every spring, and it's outcompeted mint and day lilies. Better suited to a wide open lot, it's a pretty and powerful colonizer.

In contrast, my favorite native plant has to be wild ginger. The root is small but mildly gingery, and it can be harvested for some of that flavor if you have enough of the plants to sacrifice a few. I was introduced to it by my partner, who took me and a few other volunteers on a plant rescue. He does environmental restoration work, and his project was going to disturb the earth around a stream in order to redirect its flow and stabilize it. We dug out the flourishing native plants and relocated them out of harm's way. I brought home a ginger plant. I thought maybe I could keep it in a pot on my porch, but soon a squirrel discovered it and tipped the pot over. I decided it deserved better treatment than that.

In honor of my partner's mother, whose name is Ginger, I planted this little treasure in my flowerbed on Mother's Day. It went into an area that I had cleaned out, previously filled with broken glass, construction rubble, cigarette butts, bottle caps, and other debris. Because this was my first native plant, I sort of expected it to just wilt and die in the poor urban soil. It has surprised me year after year, returning a little more robust each spring. Not aggressive at all, just minding its business in the shade, where forgotten trash once littered. Its velvety round leaves invite stroking, and I pause to remember the forest where it came from and admire its persistent and quiet resilience.

(continued on p. 110)

Stinging nettle.

Now I live in a more suburban space, with an actual front yard and backyard, which has opened up a small world of possibilities for native landscaping. The main feature of the property is a giant pin oak, a quick-growing tree that developers installed in the neighborhood when they built the cottage-style homes in the 1950s. Now these trees are aging, requiring some professional care, and nobody seems too fond of them. They are quick at dying as well as growing, so homeowners need professional help with trimming and cutting and, in some cases, felling and removing them. My partner and I enjoy watching all the life and diversity that this single tree supports. Under its shade sprawls another beloved wild ginger, jack-in-the-pulpits, wood poppies, foam flowers, stinging nettles, and some ferns and sedges.

Oak transplants.

High plains grasses.

I particularly like the nettles, which I harvest with gloves and scissors, trimming but not destroying the plants. I boil and eat the greens and drink the tea (the stinging effect of the tiny hairs on the leaves is neutralized by cooking or drying). Nettles are high in vitamin A and iron, and their anti-inflammatory qualities are perfectly timed with allergy season. Some nettle tea with local honey is my staple drink in the spring and fall.

Pin oaks produce an abundance of tiny acorns, their sound on the roof signaling that fall has arrived. The acorns feed the gray squirrels that keep our cats entertained for hours. Those squirrels plant the acorns all over the yard, of course, and forget many of them. Instead of mowing down the tree seedlings that sprout up, my partner transplants them. After potting them up, he takes them out and plants them in some of the larger landscape restoration worksites. I wonder if, maybe by adding the genetic diversity of pin oaks from the city to the wilder ecosystems, we can help strengthen their chances of survival, too.

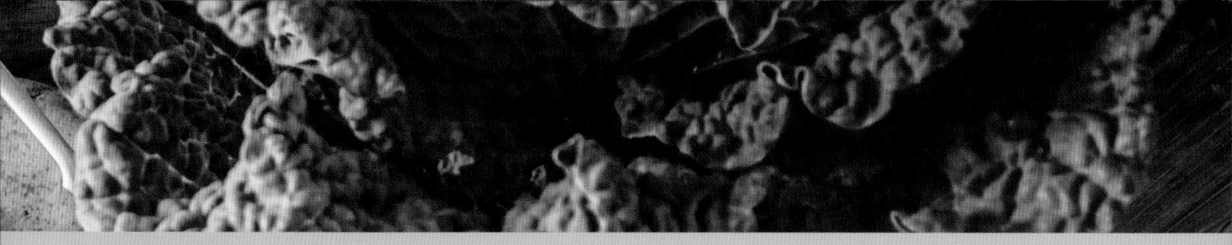

Lexington, Kentucky, has been a wonderful place to learn about native plants, while I admit that certain plants remind me of my roots in a more arid environment. Living on the high plains of Texas, prairies full of wildflowers and waving grasses decorated the monotonous horizons against a backdrop of infinite skies.

I rediscovered those grasses here in the Bluegrass State. This nickname might come from the little bluestem, a beautiful bunch grass with pale blue stalks in the spring that mature into a vibrant orange-red with wispy white seeds in the fall. Prairie dropseed also excites my senses with its intoxicatingly fresh smell. The scent of prairie dropseed cleanses my head and reminds me of the change in the air before a rain or the smell of clothesline-dried sheets.

I could list a number of great native plants and all the ways they benefit wildlife, how you can use them for food and medicine, and when and how you should plant them. But the best advice I can give is to connect with native plant lovers and let them show you the way. Right now, I'm going to go dig a hole to transplant a native columbine I got at a plant exchange. It will join another little columbine, and both of them remind me of my days in the mountains of Colorado, where the columbine is the state flower. I hope its flowers attract a few hummingbirds, my mother's favorite bird. Go out and find your own reasons to connect with native plants. Plant a few, and see how they grow on you. Embrace your inner wild.

Colorado wild columbine.

WE CAN DO IT

Know-how. Can-do. Common sense. Muscle memory. Born to do it. In my genes. In my blood. These words sound like what home improvement centers and 4x4 truck dealerships use to pump up their commercials. It's as if a prerequisite for growing food is to have been born driving a tractor and mending fences. What if you come into it later in life?

We are abundantly blessed with information. The "know" of the know-how is there. The "how" of the know-how requires something beyond the brain. I've taken courses, researched, and written about aspects of farming. However, applying the book learning to the field takes a different kind of ability.

Oral traditions passed down from one generation to the next, teaching without lecturing, working alongside, and guiding when to sow seeds, how to space rows, when to harvest, and why to leave certain weeds and pull others—the wisdom of our elders seems like voices fading into the mists of yesteryear. But that wisdom is still with us. At the 2016 Southern Sustainable Agriculture Working Group (SSAWG) conference, certain aspects and experiences brought the "how" to life for me, even midwinter from within climate-controlled conference rooms.

Mentors. New farmers are desperate for mentors. Maybe our grandparents lived on farms, and then the Baby Boomers sold the farms and moved to the cities. Now, Generation X, Generation Y, and millennials want to run the farms, and we need some help tapping into our agrarian roots. At the conference, Janie Sims Hipp spoke for the Native American youth she works with. They are motivated young farmers, many of whom work with their grandparents to raise livestock and learn the knowledge that skipped a generation. "The sooner we get young people in the driver's seat, fully equipped with some really complex ideas, the better off we'll all be," Janie expressed during the conference panel discussion. She started a native youth food and leadership summit, which is growing every year and giving her hope for the future.

Other panel discussion participants shared their stories of what has worked well over the past twenty-five years for the sustainable farming movement along with humorous stories of what has not worked so well. One story told by farmer Alex Hitt, about dealing with a flood on a dark and stormy night, stood out. It conveyed the stress and helplessness that inevitably takes a farmer by surprise when the elements of nature show their strength and a natural disaster wipes out any semblance of human-imposed order on the land.

What came out of Alex's experience was more valuable than what he could have hoped for: the understanding that it is possible for disasters to lead to the best, most profitable years. Alex reflected on his land's unexpected resilience after the traumatic test of its integrity: "We all have this faith that taking care of soil is the right approach...but we don't know for sure if it's working until it really does.... We know that we'll survive; it actually works." This type of tried-and-true, patient, long-term knowledge can only come from years of experience or years of listening to those with experience.

Organizations. One-on-one guidance is the ideal way to learn how to operate a small farm, but many new farmers don't have an elder to look up to. Formal programs for apprentices and volunteers also need help to facilitate connections between seasoned and budding growers. At the SSAWG conference, it was encouraging to hear about organizations that are supporting the beginning steps that will help newbies gain hands-on experience.

Incubator schools, such as the Roots Memphis Farm Academy (www.rootsmemphis.com), are working in urban areas to prepare the next generation of farmers. The organization's website describes the leadership that this nonprofit provides. It "trains and launches new, sustainable farmers in the Memphis region by combining sustainable-farm skills training and small-farm business management classes with an incubation process that connects graduates to land, financing, markets, and ongoing technical assistance and consultation."

Grow Food Carolina (www.coastalconservationleague.org/projects/growfood), managed by Sara Clow, presents an excellent model of a food hub that is connecting local farms with markets. This organization is part of a larger effort to conserve coastal land and recognize the vital connection between the local food economy and ecological stability. By supporting the financial viability of small farms, they give new farmers more reason to follow their dreams and give it a go.

In addition to grass-roots establishments working in their local areas to connect and support farmers, conferences create an indescribable atmosphere of acceptance and camaraderie by gathering like-minded folks. The social capital generated by the SSAWG conference and the feeling of connection with a greater community is a foundational piece for beginning farmers. SSAWG organizers provided ample opportunities for chatting informally with presenters, plenty of downtime between sessions, and an area designated for interns and apprentices to meet up.

Field trips. While talking, listening, learning, and sharing are really terrific, there's nothing like just getting your boots a little dirty. Many conference attendees—especially those who learn best by doing—felt that the field trips were the most worthwhile aspect for them. Even in the middle of winter, much can be learned just by walking in another farmer's footsteps.

One of the field trips offered was to Salamander Springs, a local beacon of small-scale permaculture. I had firsthand experience, as I had volunteered on this farm for a short time. Surrounded by forested hills and a clean, flowing creek, it's an ideal place to learn how to interact with your land as the holistic system that it is. An off-the-grid home tops a slope that leads down to an outdoor kitchen and a plethora of veggies, herbs, beans, and corn, along with ducks, chickens, and geese. With permaculture teacher Susana Lein welcoming greenhorns to her farm, Salamander Springs exemplifies what can be accomplished by building soil while building community.

— — — — — — — — —

As I took farm-girl baby steps in my own backyard in the spring following the conference, the flutter of anxiety over making the right decisions and sowing the right seeds at the right time began to quiet down. Together with my partner, we sorted through all the seeds I've held onto from my past few small garden efforts. Wow, there were more than I remembered! Viable or not, there was an abundance of everything—seven varieties of kale! He prepared the soil in the flats, and I overcame my overanalytical, cautious tendencies. My spirit began to hum with the forthrightness of just doing it. Yes, I read the packages to find out if it was OK to start these before the last frost, and, yes, I labeled row markers instead of scattering the seeds chaotically (which both I and my partner have enjoyed doing in the past). Yes, I wrote down how many days to maturity, and, yes, I transferred all of this information to a calendar to make sense of it later. The soil felt cool, and I broke up clods and tossed away sticks as we made knuckle-deep grooves, sprinkled in seeds, covered them, and tapped the soil gently. And then I breathed. That was refreshing. That's how you start an urban farm.

REALITIES OF FARMLAND RESTORATION

Winnie the Pooh really knew his land. If you look at a map of the fictional Hundred Acre Wood, you'll discover hidden gems such as Sandy Pit Where Roo Plays and Area of Six Pine Trees. There's also Bee Tree and Area of Big Stones and Rocks, and my personal favorite, Eeyore's Gloomy Place. It seems there are habitats for all creatures where Pooh lives. This kind of diversity isn't associated with conventional agriculture, yet many bright-eyed beginning farmers have dreams of integrating natural ecosystems and restoring wildlife habitat on their property.

Steve and Leah have created their own "hundred-acre wood." It used to be 500 acres, four generations ago, when Steve's ancestors raised beef and dairy cows and tobacco. As the family aged, they sold off the land in portions. Today, Steve and Leah are grateful to hold on to the remaining property. Farming is not their passion, but land stewardship is. In the past ten years, the couple has planted tens of thousands of hardwood trees.

Turning dusty rows of monoculture into a diverse, abundant forest requires vision, knowledge, patience, and heart. Here are some things to keep in mind if you are considering an ecological restoration project.

Start with a restoration goal in mind: What is your desired ecosystem? Imagine lush woodlands bustling with birds and butterflies... Cool streams meandering through orchards... Leafy forests shading and stabilizing tired earth... How long do you want this ecosystem to thrive? Will you convert it back to crops at some point?

Carolyn and Jacob Gahn grow sorghum to make their Sweetgrass Granola products and are raising most of their own meat and vegetables. Their ultimate goal for their family farm is a closed-loop system for self-sufficiency. Carolyn says, "We see ecosystem goals as larger than our farm—it's a whole community process that also includes a local food system, so we realize other farmers and consumers need to be involved with all of these goals."

If you know what you want, you have started on your conservation path. To make your dreams reality, set short-term, mid-term, and long-term goals. Short-term ought to include getting to know your neighbors and finding collaborators. Mid-term plans foresee the next decade and include a succession plan of introducing the right species at the right times. Long-term plans consider generations that will follow you and the legacy you will leave.

Get to know the people who can help: Give your Garden of Eden dream a reality check with help from experts. Make a checklist of resources that provide knowledge and funding, and then get ready to make some phone calls.

Robert Hoffman, wildlife biologist and restoration consultant with Roundstone Native Seed, gives this advice on who to contact: "With so many programs to choose from and many practices within each program, the best option is to go to your local USDA service center where both Farm Service Agency (FSA) and NRCS are located to get more information. After all, these will be the people you will be dealing with on your project, and they may also be able to get you in contact with local conservation programs through your state (like your state wildlife agency) or even county-level programs through your local Soil and

Water Conservation District. Most of the time, they are located in the same office."

Tyler Sanderson, a technician in biosystems engineering and forestry, explains that certain states require permits for making any alterations to farmland, whereas others will allow plowing over ephemeral streams. It's important to know the scale of your project and who else it will impact, even if you plan to implement it alone.

Depending on the goal of your project, branch out to include representatives from your state's division of water and division of forestry as well as the USGS. Less obvious support may come from special interests such as Ducks Unlimited, Trout Unlimited, or the Nature Conservancy. Don't overlook your local anglers' associations, birding clubs, native plant enthusiasts, and town governments, who have various reasons to support natural habitats.

Steve and Leah consulted with their state's department of forestry, who loaned them a tree transplanter to use with a tractor. It enabled them to plant 500–1,000 trees per acre, which they managed to do on weekends with a few extra helpers. Jacob and Carolyn completed a Timber Stand Improvement project with the Kentucky Department of Forestry. Carolyn explains the benefits: "This program was started to provide landowners with an income stream from their woods without having to clear cut. The goal

(continued on p. 116)

is that it creates a long-term stewardship plan for your woods. It also worked for us because we were able to cut our firewood for the next two years."

Understand your land and water: To learn how your soil fits into a larger picture, and how it affects your watershed, get a soil map from the USGS or use the Web Soil Survey, a user-friendly website that shares soil data from nearly every county in the United States and creates a customized report of your area. Some states have regional watershed coordinators who can work with you to understand the impact your projects will have downstream as well as who is upstream from you and impacting your water quality.

Inventory all of the problems on your land, even if they aren't obviously connected to the area you want to restore. A conservation technician can help you identify inconspicuous assets. For example, you can cut and run invasive trees through a chipper to make mulch for fruit trees or garden paths. This clears out space for native plants to have a fighting chance while recycling the carbon energy captured in the woody material.

Ben Leffew manages the nature preserve at Shaker Village, a unique property that includes a historic landmark, living museum, and working farm. Among the many interests that Shaker Village works to balance, they have restored 1,200 acres of native prairie and lease 750 acres of farmland and pasture. When the restoration project began in 2009, Ben worked with the Kentucky Division of Fish and Wildlife, which was focusing on bobwhite quail habitat. Ben explains, "We take a bottom-up approach...build the habitat and manage for pollinators and native grass, and the birds will show up." Along fencerows and in marginal areas, native habitat for grassland songbirds is flourishing. Birders are also flocking to the site, a great ecotourism boost.

It might get worse before gets better: Going native isn't going to be pretty. Tyler Sanderson warns landowners to brace themselves for criticism. Even country neighbors expect neatness, and natural habitats can look woolly and scraggly as they grow into their own beauty. Several treatments of professionally applied herbicides and controlled burns may be required to bring an end to invasive species.

Robert Hoffman warns that the most common mistake landowners make is not adequately preparing their seedbed. He says, "Without giving proper consideration to the establishment of your project, it may be doomed to fail. This can be a reality check for some landowners because although you may decide this month that you want to plant, it may take one to even two years before planting should begin."

Steve and Leah can vouch for the transition period's stretching out longer than expected. Steve points out a simple fact about trees: "They really grow slow!" The diverse tree species they selected, including ash, oak, hickory, and fruit trees, mature at different rates. Ten years into their forest project, the tallest trees reached around thirty feet and required pruning. Leah is happy to report, "Now they're tall enough that I can ride the horses down through the trees!"

Take care of the land, and the land will take care of you: Your long-term goals should include the inheritors of your project. If you have children who will take over, make sure that they are involved now. Results will take time, especially if you start a reforestation project. As Steve reflected, "Three generations before us tried to clear trees off, and here we are trying to plant them again. We are trying to look seven generations down the road."

The most important thing you can do with your land is to use it for education. Tyler Sanderson points out, "People want their water to be clean, but they don't understand that mowing up to the edge of their stream is bad."

Teaching others can take many forms. Shaker Village allows hunting and hosts tours of their restoration projects. You don't have to run an agritourism business to be a model for other landowners. Simply invite groups of friends over for work parties or volunteer to host an agricultural field day through your extension office. After all, even Winnie the Pooh's Hundred Acre Wood still has a Floody Place. Maybe Tigger or Christopher Robin could help with stabilizing his streambank.

Final words of advice: Carolyn and Jacob advise, "We would say do one at a time. There are multiple NRCS and conservation programs that each farm can do, and it sounds great on paper to sign up for all of them. But these are contracts that you are signing to get the project done by a certain time, and if you don't complete them, there is a penalty. Use these programs as a supplement to your farm income and focus on building your business."

From Steve and Leah: "We're glad to accomplish what we did, but now we're thinking we could harvest blueberries sooner than hardwood trees. It might be nice to help ourselves out a little. Be selective in what you grow. Make sure you can harvest what you grow."

GREENHOUSES: GET GROWING

When you walk into a greenhouse, the "green" is what you notice first, but the "house," the hardscape, is supporting the life and shaping your experience in the space. Every structure serves a purpose and interacts with all the other elements on your site, whether you pay attention to it or not. In permaculture, the intention you put into the elements shapes how they relate. For best results, design your own greenhouse or have a permaculture expert help you customize a plan. Every greenhouse owner should take the time to study basic solar principles, such as the orientation and angle of surfaces to maximize radiation and the laws of thermal dynamics. Skimming over the research and just buying a prefabricated greenhouse-in-a-box could lead to using more fuel just to heat it. The main goal is to let nature do most of the work and to disconnect your production from dependence on nonrenewable inputs.

Permaculture design always begins with considerable site observation and analysis. For a greenhouse, the idea is to harvest the sun's energy when it is most abundant and hold on to it for as long as possible. Knowing how the sun acts is the first step.

In *The Food and Heat Producing Solar Greenhouse*, Bill Yanda and Rick Fisher give detailed instructions on how to observe the sun's patterns, wind flow, and obstructions, as well as drainage, building codes, and greenhouse geometrics. Observation of patterns informs decisions about the details, such as orientation of the greenhouse. "Because the days are so short from October to March, both the plants and the heat-storage features of the greenhouse need every available minute of sunlight. Designing in accordance with sun movement patterns gives the solar greenhouse automatic advantages over conventional greenhouses for winter heating and summer cooling," say Bill and Rick in their book.

The north side of the greenhouse should be as protected as possible and can even be built as a partially subterranean wall or connected to your home, chicken coop, or other buildings that you'd like to keep warm. Jerome Osentowski's Phoenix is attached to a woodworking shop that receives a surplus of warm air through a vent.

The opposite is true as well: connecting a renewable heat source to the greenhouse can make life easier when the winter chill really takes hold. Phoenix and its workers enjoy a wood-burning sauna on the coldest of nights, a backup system that provides not only a welcome heat retreat but also a socializing space to host friends. When you add aromatic herbs, such as rosemary and eucalyptus, to a steaming pot of water, it helps draw out toxins as you sweat through the winter solstice.

Over many years, Jerome has created a sustainable food-producing wonderland full of papaya, avocado, and citrus that is also a great place to hang out. The greenhouse roof supports a sunny deck, a relaxing space for sipping tea or doing yoga. Inside Phoenix, Jerome's favorite place to conjure up designs is in his hammock, which hangs below guava vines. A sleeping loft is nestled above banana leaves, and on some nights his dreams are infused with the scent of night-blooming jasmine.

HOLD YOUR HEAT

Permaculture is all about balancing and using elements that work more than one way, and thermal mass regulates temperature despite the extremes of the outside weather. As defined by Bill and Rick, thermal mass is "a mass of material in the greenhouse that soaks up heat from the sun during the day, then slowly releases that stored-up energy back to the greenhouse or to the adjoining structure at night." The variety of ways to hold heat is only limited by your ingenuity and access to materials. Diversifying not only helps the crops you grow, it's also a good idea when selecting structures to act as thermal mass.

Examples include gabion walls, which are loose rocks contained in wire that can provide a trellis for heat-loving vines, such as grapes or kiwi. Basic water tanks painted black retain heat better than light-colored tanks. Taking that idea to another level could mean adding an aquaponics system in which you farm fresh tilapia and recycle their nutrient-dense waste into plant fertilizer. Working on the simplest ground-level idea, you could utilize the space under your plant tables as worm farms, harvesting warmth and red wigglers to share in other places.

Through their business, Ecosystems Design, Inc., Jerome Osentowski and his partner, Michael Thompson, have designed a subterranean heating and cooling system that they call a *climate battery*. The system basically takes the ambient air in the greenhouse and, instead of venting it out, draws it into the soil with fans and ducts. Thermostats control the flow of the air, and the warm air that was stored away in the earth all day flows back up and out to heat up the cool night air.

Looping the system back to the attached rooms optimizes the flow of energy and reduces waste. Phoenix's floor radiates delicious warmth for your feet and the plants' roots. Walking barefoot through lemongrass in the middle of winter doesn't have to be just a dream.

Chapter 3

Taking Care of Yourself

Because women who grow plants and raise animals are the nourishing healers of their families and their interwoven communities and environment, it is worth dedicating a section of this book to finding support for self-care and replenishing our own wells. In this chapter, we will take time to explore how you connect with your body's way of doing your farming work and how you honor nature's cycles within and outside of yourself.

"Our needs support us thriving." A handout distributed by Corinna Wood of the Southeast Wise Women displayed this message with a circular map of the moon's phases and their correlations within our bodies as well as with the work we do outwardly in the world. Corinna is a leader in the Wise Women tradition, which offers an annual conference and instruction in herbal medicine. A focus of their teachings is self-love because our personal healing is central to our ability to heal others or the earth.

Susun Weed outlines three traditions of healing—scientific, heroic, and wise woman—in her book *Healing Wise*. The scientific tradition began around 1,500 AD, the heroic tradition around 1,000 BC, and the wise woman tradition around 50,000 BC. In the scientific tradition, the healer is a mechanic, using treatments involving drugs and surgery to fix isolated problems and trusting test results to prove the enemy has been beaten. In the heroic tradition, the healer is a ruler who saves or punishes the patient by purging toxins and cleansing the spirit's physical temple, the dirty body. In the wise woman tradition, the healer views the body as perfect and complete and provides compassionate care that focuses on support and nourishment; this tradition

Which tradition of healing do you practice for yourself? Describe your self-care.

Which tradition of healing is predominant in your household? Describe.

Describe what attracts you to your chosen tradition of healing.

views disease as an ally for transformation, working through unique and unimagined variations.

We've all been there. Rushing to meet a deadline, pulling an all-nighter, lifting heavier things than we should, pushing ourselves to meet too many demands. These are opportunities to find our limits, recognize them, and learn from them. Boundaries are important. They create the form that balances our freedom. Fence lines tell neighbors who owns which plot of land. Schedules determine what work is to be done that day. Garden beds separate food from pathways. Saying *no* allows us to say *yes*.

The choices we make to take care of our land and ourselves, our families and our communities, all reflect back on our priorities, values, and goals. Sustaining ourselves is just as important, if not more so, than meeting anyone else's needs. Some women need to be reminded of this frequently. Breaking free from oppressive attitudes about what a woman's place is can bring a deluge of opportunities and liberties. We may not have been taught how to discern where our limitations are and how to successfully find balance in life. The irony is that healthy women's bodies are crucial for the health of the environment and successive generations.

Women's bodies serve as clear indicators of environmental health. Americans seem to be a thriving population of robust, hearty humans, with modern medicine replacing worn-out parts and propping us up to live longer lives. But what about the abundance of disease? Cancer, heart disease, and diabetes, for example, with much of it stemming from obesity. How do we rectify the disparities of malnutrition in a throwaway society, where convenience stores and pharmacies alike are filled with aisles upon aisles of empty calories? How many women do you know who have been affected by breast cancer? How many do you know with thyroid issues? How many more experience vague symptoms that doctors just can't diagnose? Could our breast tissue and endocrine systems be like canaries in coal mines, sending us grave warnings that all is not well in this world? It's time we take a cue from the natural world and get in touch with our bodies, our emotions, and our limits, and find balance.

Even if we feel influenced by a testosterone-driven work environment, women can take the lead in rational behavior to sustain self-care. Angela Wartes-Kahl lives and

works with a mix of men and women farmers of various ages and abilities. She says they all work together well, and generally the mix has been more female than male. She likes the ratio of two men for every three women. Why? "Just to mellow out the 'git 'er done' attitude," Angela says. "Instead of just 'Let's go out and do it,' it's 'Let's go out and think about how to do it right.'"

Angela remembers times when that "get 'er done" attitude prevailed and lessons were learned. "We had a couple of summers when we were building a barn, just powering through projects to beat the weather or…to get them done, and people got hurt. It was good to prove to ourselves that we were able to do it if we needed to, [that] we had the strength, ability, and perseverance to get through it. In retrospect, it's easier and safer in the long run to think through our decisions and make good decisions. We don't need to push this hard. Over the years, we try to scale back our exuberance over projects we can get done in a certain time period. We try to be more realistic and not push ourselves as hard."

KNOW YOUR BODY'S WAY

Just as establishing our boundaries and respecting our limits help us create form in which we have freedom to work, the same is true for understanding our bodies and honoring our bodies' needs. Learn the proper structure and form so that you can express yourself and work freely, without injury.

PROACTIVE PREVENTION

Lisa Munniksma has worked at farmers' markets alone many times, loading and unloading, setting up produce and displays, and she notices that physical strength is a big issue. She is not usually quick to ask for help, often having an "I-can-do-it-myself" attitude. She is solution oriented and finds workarounds despite working with others who aren't consciously considering the ability of the person who will be unpacking their products.

She gave the example of pulling up the farm truck to set up at a farmers' market. "There was one really large cooler full of meat rather than two or more smaller coolers that one person could reasonably lift. It's not even from a woman versus man standpoint—it's just knowing that you don't have to have two people to unload."

Using our big brains—those tools we have with us all the time—and our ability to communicate about our needs will help in these instances. Remind coworkers that we'd appreciate more loads that are lighter instead of fewer loads that are heavier, for example. Help each other foresee human needs and speak up about how they can be met.

Another way to save some wear and tear on your body—while also keeping a sense of order about your work—is advance planning. I naturally tend to run around, forgetting where I left things and then backtracking to find them, or starting one project and switching to another when I realize there's something more urgent that needs my attention. Watching a proactive community garden coordinator deal with this common tendency toward chaos helped me learn a couple of ridiculously simple techniques.

1. **Take notes.** Before I do anything, I walk through the space with a notepad to see and jot down what work I need to do. For a garden, my list could look something like this: weed lettuce bed, thin beets, water everything, mulch raspberries, clean up used potting trays.

2. **Know my time limit.** If I have one hour, I look at my list and consider which task or tasks I can complete before I have to leave the garden. If I have someone helping me, we decide whether we will divide one task or work on two separate ones. The more familiar I get with the space I am working in, the more accurately I can estimate the time I need to get the work done.

3. **Plan my trips.** Gardening sustainably, as far as my body is concerned, is often an exercise in trip planning. I visualize the route I am going to take, where my materials are, and how many times I will need to go back and forth, and then I get to work. Every time I walk back to my water in the shade, or my tool shed, or the compost bin, I consider whether I can take something with me. Fewer empty-handed trips waste less energy and keep my head in the game, too.

Pay attention to your movements as you bend, stretch, lift, and otherwise exert yourself.

There's a popular children's song called "Head, Shoulders, Knees, and Toes." You tap each body part while you sing it. The idea of the song is to help children learn those body parts and their names as they are becoming aware of themselves. As adults, we can use this song as a gentle reminder to keep our backs straight: the head over the shoulders, and the shoulders over the hips. When we bend our knees to lift something, our knees should be over our toes, but not beyond them.

It's sound advice to pace yourself as you do manual labor so that your parts keep on working as long as you need them to. Just like the parts of a machine, the parts of your body need maintenance to stay in working order. Staying hydrated and avoiding inflammation provides lubrication at the points of friction, and paying attention to your weaker spots by building up the muscles around them are a couple of basic principles for long-term health.

It's hard to discern fads from truth when it comes to health, but looking at our ancestors' diets and their behaviors can clue us in. Your intuition is a sage medical advisor as well. Take a moment to reflect on your family's health history and diets.

In Nia, a holistic movement form I study and teach, we learn about the body's way and your individual body's way. Every body is composed of pretty much the same

Who has lived the longest and/or healthiest life in your family?
What types of food did/do they eat? What were/are their daily activities?
What kind of spiritual life did/do they have?

Who has lived the shortest and/or unhealthiest life in your family?
What types of food did/do they eat? What were/are their daily activities?
What kind of spiritual life did/do they have?

What foods help you feel your best?

What activities help you feel your best?

structural parts: 13 joints, 206 bones, a pelvis, ribs, a head, and all of the smaller parts that are integrated within that system. Learning how that system is designed to move is one aspect of self-care that you can find through any number of movement and exercise modalities. Learning *your* body's way is something that only you can find.

I've found some holistic approaches—including yoga, tai chi, belly dancing, tango, Pilates, hula hooping, and the fusion of several fun and fluid techniques within Nia—useful for understanding the body's way so that I can get in touch with my own body's way. To relieve tension or aches that I haven't been able to prevent or avoid, I see a chiropractor and get massages on a regular basis. This is my current self-care practice, and it is subject to change as my body's way changes.

I also use a soft foam roller and balls to massage and hydrate my fascia (connective tissue) through the Melt method. Shayne Wigglesworth teaches foam-rolling workshops and explains how it works: "Foam rolling is like a massage but self-care and not as expensive. It supports what your massage therapist is doing. Over time, misuse and habitual use form scar tissue that gets hard. Arthritis occurs from repetitive use when tissue breaks down and there's less fluid in the joints. Foam rolling breaks up scar tissue and allows fluid to move back into muscles and joints."

Shayne also teaches Nia and practices every morning before her baby wakes up. She believes that "in order to function properly every day, you have to take care of the vessel." She knows from experience that forcing the body to go beyond its limits is counterproductive. "I was a collegiate athlete, did step aerobics, and lived a 'no pain, no

WHAT IS THE NIA TECHNIQUE?

The Nia Technique originated in 1983 and is the original fusion-fitness program in America, based on healing arts, martial arts, and dance arts. It is mindful, rhythmic dance practiced barefoot to eclectic world music. Nia encourages every person to get in touch with their senses and adapt movements to fit their personal needs. As a workout, Nia is non-impact and focuses on the balancing and healing aspects of exercise through understanding the body's way. As a holistic self-care practice, Nia classes are designed to support fitness in the mind, spirit, and emotions, as well as in the body.

Yoga, tai chi, and other practices help build body awareness.

gain' lifestyle. Nia taught me I can live in great shape without being sore, and I can lift weight as my own body weight. I've learned how to use my body more ergonomically, how it's made to move, how to move more sustainably. This is something I can do until I'm very old, because I use my body in the way it's meant to be used."

Shayne worked as a manager at a natural foods co-op for over a decade and promoted locally grown, organic food throughout her community. She understands that a new holistic health craze pops up every day, and she encourages people to find what works for them. Her Nia dance classes are about relationships, which is different than many exercise regimens. "It's a great community of people building lifelong friendships. It's not like going to a gym [where] we work and leave. It's a different sort of connection."

JUST BREATHE

To keep the body capable of doing fun farm work, cultivate an awareness of your breath and form throughout all your activities. Remy Hendrych identifies some simple indicators: "There are some really basic positions that the human body is supposed to be able to do…kneel, sit in a squat, and bend over with no pain or loss of breath. There are a gazillion things you can do to help these basic human postures…during farming, but you may not do them with consciousness, with attention. That sometimes leads to exacerbation of whatever pain is already there."

Remy has a varied background as a fitness professional. In addition to urban farming, she is a teacher and cofounder of Urban Ninja Project, a gym that teaches natural movement. She echoes the wisdom from the yogic tradition, paying attention to the body's subtle energy and listening to the inner messages. She says, "I found the most pleasure and progress just in getting into a posture, holding it, and moving around in it." Remy makes an effort to apply that discipline to her gardening practice. "Sometimes I'll kneel and catch myself in a position [where] I can't breathe very well, so what should I do—stop my whole day and do yoga or whatever current fitness fad is going to fix this thing? No. What I do is a combination of things I've learned from physical therapy and animal flow and other modalities; it's simple. Just stop and listen, and ask your body, 'What would you like to do?' Maybe I'll sit down and get in a position where I can actually breathe and listen, take deep breaths. Usually the answer is, 'I want to keep going, but I want to go more slowly or more consciously.'"

Conscious awareness is the key. It is similar to, or perhaps actually is, a moving meditation, watching the breath while you move as effortlessly as possible. Remy recognizes that. She says, "When I get into those places of pain or discomfort, it's usually because I'm not paying attention to my breath or the actual thing I'm doing—

Remy's background in fitness and movement serves her well on her urban farm.

my mind has wandered." For example, "If I go back to 'I was picking hyacinth bugs off this cabbage' and thinking about [everything else] I was supposed to do, what he said and she said and all this other crap, then I can go back to being present with the hyacinth bugs and going slower, getting in a different position, being mindful of what I am doing. That usually leads to [being more attentive] the next time I do it." Over time, she has become much less likely, when doing farm tasks or any physical tasks, to let her mind wander.

When your mind is on what you are doing, you're more centered in general and make better decisions. Remy summarizes, "The more mindful I am, the more likely I am to make the choice, that morning when I wake up, to go do yoga instead of getting on the computer or doing some menial task I could do later. I'm more likely to make a good physical choice for myself when I'm more centered in general."

At Soma Ranch, as the name implies, everything they do is based on somatics, which is a method of moving the body's way. Helen Terry and her husband, Joe, aim to serve a rural population in need of repairing damage done to their bodies through years of manual labor. Helen says, "It's a complement for anyone who works in a farm environment, where chores tend to be repetitive, overusing certain parts of the body. Somatic movement brings in systemic movement, which is moving all your thirteen joints, balance between the front and back of the body, and balance between dynamic

movement and ease. The body thrives on dynamic ease." Helen explains that our bodies strive for balancing yin and yang. She says, "Farming can be more yang, so bring yin to that, and the body itself reveals the body's way."

Something I enjoy about tai chi and Nia is that all the parts of the body have a chance to do something rather than just the parts we force into labor over and over. We have the opportunity to bring more yang to the parts that are atrophied from underuse, and more yin to the parts that have been overworked. As Helen says, "If we have something like Nia to balance it, we can do tasks more optimally and with less pain and more ease. The goal is to be able to be more effective while you're working but have more fun when you're playing, too."

How about you? What have you learned about your body's way, and how did you learn it?

What are the modalities you currently use to connect with your body's way?

What resources do you have to connect with your body's way?

HEALING

Jessica Ballard works in a farming environment that is run by women and relies primarily on female workers who are recovering from intimate partner abuse situations. To practice her own self-care is critical for Jessica to maintain balance with the drama, hormones, family challenges, and hard labor that her job calls for.

I asked her about her self-care practice, which is centered on creating space for her own time every morning. She explains, "I do a little yoga and meditation, my spiritual practice, some prayers every morning, and get myself grounded for the day." She uses astrology to understand the ebbs and flows of activity and personalities that transition through the farm on an erratic basis. She says it allows her to give herself "grace, or charge forward," depending on the day. She doesn't claim any single religion but is well-read in a diverse array of spiritual practices. "I don't really have a religion but pray all the time," which is evident by spending any time with her among the flowers she grows.

Remy's urban garden features spots for relaxing and observing nature.

CYCLES

Remy Hendrych, the health coach and permaculturalist, is also a self-described garden fairy. When I asked her about the path that led her to establishing an intentional community and urban garden, she taught me a word: *cyclicality*. She said, "Cyclicality forced me—it took me by the head and slammed me down, like, 'You must pay attention to cycles!'" What she means is that she renewed her health by paying attention to her body's needs and cycles. She lives with Crohn's disease and says, "I guess that's the blessing of an autoimmune disease. I was physically forced to pay more attention. A lot of people go through life not having to pay much attention to their bodies." Her diet is centered on wholesome, local foods, and her lifestyle reflects her values of nurturing the community through a connection with nature.

WHAT I LEARNED THIS SUMMER

A modified version of a blog post written by the author for Grow Appalachia (https://growappalachia.berea.edu) and published on August 20, 2012

What brought me to Watershed Farm (now GreenHouse17)? A very different reason than most of the women here—I came to Lexington to be closer to a good man. But my story echoes their need of support through a transition. It has been a big shift for me to approach life with the intention to put down roots in this city, coming from a semi-migratory lifestyle that included living in small towns surrounded by magnificent wild spaces. Now I'm in the urban jungle, awake at all hours but to siren howls instead of coyotes, jackhammers instead of woodpeckers.

I knew I wanted to find a balance between concrete and nature, as well as balance between earning a living and producing my own food. That's when Jessica and Watershed Farm entered the picture as an answered prayer. I volunteered a few hours each week during the summer in exchange for a few veggies and flowers.

Why do I volunteer? For the soul nourishment that comes from slowing down my pace, kneeling with my hands in the earth, inhaling the indescribable freshness and realness that surrounds me and grounds me, sensing the power of creation in each seed, and noticing the subtle changes from one week to the next.

What do I actually do on the farm? It's always a surprise, and I like that. I could show up and be the lone weeder, spending two solitary hours untangling morning glory vines from sweet potatoes. Or I could be swallowed up in the energy of forty other volunteers, swarming through the green rows, spreading straw mulch and clearing pathways. Often, I get individual attention from the garden guru, Jessica, who infuses the atmosphere with her infectious joy. She may hand me a flat of seedlings, rattle off flowers' names that sound like Greek goddesses, and show me how to handle them as I transplant them. Each step in the process is meaningful and rewarding, but the best part might just be harvesting. I'll never forget the day I learned how to pull a potato out of the ground without harming the plant. It's a delicate and strange feeling to blindly reach inside the earth, wiggle fingers around and detect a hard little ball, carefully separate it from the tendril that attaches it to the mother plant, and resurface with a golden (or russet) prize in your fist.

This summer has been full of firsts for me. One thing I've learned through my volunteer farming is that being a beginner is a wonderful way to exist in this world. Humble farmers know that they are never really in charge, and if you stay open to listening to nature's messages, the right teachers will come to you.

Miranda Sparks grew up on a farm and has become a professional healer. Her credentials include becoming a Certified Natural Health Professional, Reiki Master, Licensed Nurse, and Master Herbalist. Her career path is directly connected to her childhood. She shares, "Spending my life in nature has brought a sense of connectedness to the earth, which I cherish. My attempt to grow long, pretty nails has been futile due to my inability to don gloves for earth digging. I need to feel the earth on my skin. Gardening is what soothes my soul, balances my energy, and clears my mind."

For Miranda, gardening meets a much deeper need than simply producing food. She says, "Plants offer much medicine in the physical sense but also vibrationally. They are living just as we are. If we slow down and pay attention, we can hear their song."

There are so many ways that gardening and farming reveal areas that need our personal healing attention, and there are so many ways to provide healing balms for those wounds. A major aspect of Jessica Ballard's work at Greenhouse 17 is growing flowers and herbs for market. Women recovering from partner abuse help out as a part-time job and are paid a small amount to help them meet their own needs.

"It started out as a business," Jessica explains, "and we were able to get grant funding to hire women to work on the farm, as an opportunity to engage them in nature." Not everyone at the shelter wants to do it, but it is an open invitation if they are interested. "There's no expectation that someone who's been through a horrific ordeal will want to work in 95-degree heat." It's a "hook" for some, a way to lure them out into creative and thriving spaces.

Jessica speaks about some different aspects of gardening with trauma-informed care and compensating her garden help with a small stipend. "It's not only nature and beauty and being outside and what that feels like on your body and your spirit, but you can also make some money to buy gas, get a driver's license, meet a need for a small thing. Nature seeps in, and many end up loving it." Some of her clients are drawn outside to be in the fresh air; others are looking for consistent work that they can see the results of. It's not always easy to work with inconsistent employees, all of whom are working toward independence and fighting untold battles.

"Some are resistant at first. We see their barriers and sabotaging behavior," Jessica says with a smile, "but we're just hangin', and the farm rolls on. It's neat when they realize the earth's still here, the farm's still here, we keep going." She adds, "Some women have come back multiple times and seek solace in the farm, along with the shelter itself—but the farm is a sacred space for some folks."

The products themselves are healing products. GreenHouse17 grows flowers for its CSA members, weddings, holidays, and wholesale customers. Their herbs are the main ingredients in their value-added line of products called Handmade by Survivors. The shelter's nature-based programming infuses everything they do. Even clients who never come out to the farm can participate in preparing the lip balms, soaps, bath salts, and candles. Jessica shares, "It's another way of meeting people where they are—by using herbs, fresh flower petals, essential oils. It's a way to use that creative process that is empowering. You've created something that's of value to yourself or to others."

Melissa Calhoun grows medicine on her five acres. In her house, there are three shelves full of tinctures and dried herbs that she has used, will use, and will share. She says, "I typically harvest something because somebody else needs it." In addition to her own space, Melissa enjoys the wild lands surrounding her rugged herb farm. Exploration and wild foraging out there has been another form of healing through gardening. Melissa had a dog with cancer, and Melissa knew that her dog would pass soon. She openly shares their deep connection: "I asked her what she wanted to do. Just walk in the woods, once or twice a day. That was the best part of her passing, not just having that sweet time with her—her getting me to walk around places on the sixty acres we hadn't seen."

Melissa cherishes her access to a forest full of wild plants and animals. She says, "The ability to observe the land over time is so rare, even for people who get outside. Maybe that's why people have their favorite hiking trails, but even those are maintained. Here, one year it may be garlic mustard, the next year bluebells." She pauses as she recalls something she saw recently on a walk in the woods. After our conversation, I look up the herb she mentioned, *Prunella vulgaris* also called "self-heal." It is considered a weed by many, and herbalists are just beginning to appreciate its healing qualities. The properties now recognized include treating inflammation and swelling, strengthening the kidneys, and, of all things, fighting cancer.

When I asked Jessica if there's anything uncomfortable about asking women who have recently experienced an abusive relationship to arrange flowers for weddings, she says that she wonders that, too. "I always ask clients what they think about this. It's an odd, yin-yang thing. They pretty much say that they think it's beautiful, that it's cool to do this for someone else, to make an offering."

Of course, it is an optional job, and Jessica explains that it's not about talking about wedding planning to women who are getting out of abusive relationships. It's a small part of the overall work they do. The temporary aspect of flowers in bloom also resonates with the transitional nature of the shelter and the farm's role. "The amount of time people work out there is short; rarely will they plant something that they will watch and harvest themselves. All are doing work for future people who are engaged in the farm." To some, that's a healing motive in itself.

In understanding ourselves, our needs, and our limitations, we can look to nature and see that everything functions in cycles. Seasons of the year, rotations around the sun, phases of the moon, death to decay to birth—the regular recurrence of phenomena is the natural order of everything. Women's cycles are powerful because we are connecting with those natural rhythms that transcend artificial manipulation. Our lives move from infancy to adolescence to adulthood to elderhood to dying. During the time between adolescence and elderhood, our monthly cycles work to balance our bodies' essential functions. Unfortunately, women's cycles have almost become a curse word among the mainstream media, shaming us into diminishing the power that our natural changes hold. We can choose to learn from our bodies' wisdom, embrace the changes, and find our strengths in the various ways our needs and instincts support us thriving.

Jessica Godino, an acupuncturist who specializes in endocrine issues, explains how to find your flow by tuning in to the technical and mystical qualities of the monthly cycle. In a presentation at an herbalist conference, she told the audience, "Every hormone in the body is in a dance, communicating constantly, and they are all connected." To simplify matters, there are two main phases that the adult female body goes through each month: the follicular and the luteal. The follicular is the baby-making phase, which causes estrogen to plump up everything—softer skin, thicker hair, fuller breasts—while preparing for conception. Around ten or so days later, the luteal phase kicks in, releasing the baby-protecting hormones, such as progesterone.

Dr. Rebecca Booth wrote a book called *The Venus Week*, which characterizes these cycles as Venus (the goddess of love, sensuality, and erotic love) and Minerva (the goddess of wisdom, medicine, the arts, and war). Testosterone, estrogen, and progesterone combine, swell, recede, and reset at regular intervals throughout the month, and tapping into these shifts can bring us deeper into our innate power to work with what nature has gifted us.

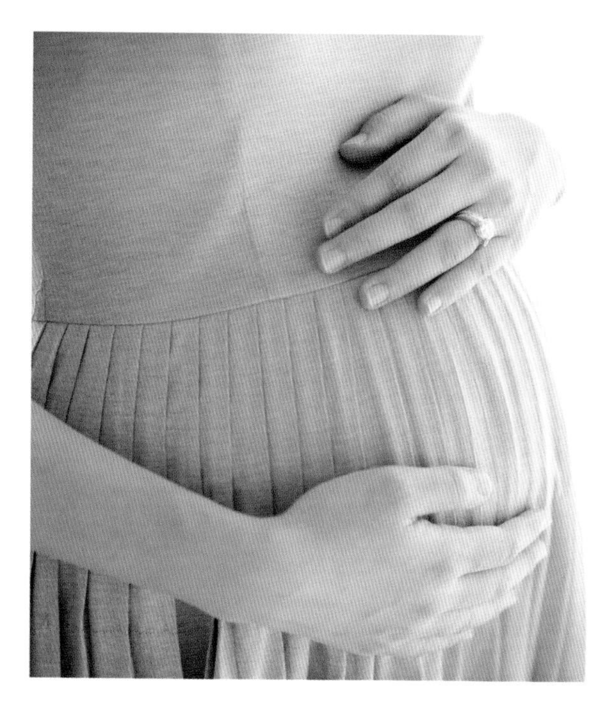

The larger cycles of a woman's life, moving from child-rearing age to what is affectionately known as "crone" age, brings major shifts in the energies that influence our needs and decisions. Once menstruation stops, the hormones don't just dry up. Instead of producing the baby-making resources, that energy is converted to all sorts of different outlets.

This is a wonderful time of life to embrace the hobbies and passions that you set aside while you raised a family. I know several women farmers who waited until retirement to give their hobby their full attention. Pamla Wood, for example, is sixty-four years old and has a forty-acre farm with wildlife, sheep, goats, hens, and vegetables. She worked at state agencies in soil conservation and water quality. Her networks introduced her to farm education opportunities, and when she was ready to begin farming in her fifties, she says, "I pretty much had all the help I could use, from and through Kentucky State University, other women farmers, a small grant from the local Conservation District, University of Kentucky, and Kentucky Women in Agriculture."

Pamla is proof that many things grow better over time. She sees this as a strength carried by many women farmers. "Women are generally more patient, in my opinion, in letting projects develop; for example, many take their time to learn how to take care of livestock and are not quick to give up on a new project."

TRANSPLANT SHOCK

Excerpted From a June 2012 post on the author's personal blog

I started my eastward migration five weeks ago. Week 1: driving. Week 2: getting settled in, dealing with lingering obligations, and doing some light farm work. Week 3: real farm work! Week 4: sick, dilemma to leave farm, finding new home. Week 5: recovering.

That one week of real farm work taught me a lot. On the first day, I was happy to begin a rigorous routine as I helped Susana shovel horse manure, which I actually enjoyed! It's a great workout for your midsection (I highly recommend it), and something about cleaning a space for a neighbor in exchange for a truckload of nutrients for the soil is gratifying. In the afternoon, I sat in the shade and transplanted tomato seedlings into separate containers. Looking back, this also taught me something about myself.

Under Susana's careful guidance, I put one finger on each side of the seedling and jiggled the moist soil of the tray where they had spent their formative first weeks. I gently loosened their roots out of the shallow nursery and tried not to tear any. I then added a little biochar to cover the bottom of each one's new (recycled) single bed and added a handful of a rich potting mix to the container. Next, I set the tiny root ball of the plant onto the fluffy darkness, added a little more soil on top, carefully swiped back inside any escaping root hairs, and then tapped it all down gently.

The author's successful tomato transplants.

Ah! New home. Now that dozens of precious little tomato plants of heirloom varieties were placed in their new dwellings, lined up in a grid, I kept them in the shade and lightly sprinkled water over their tops. I practiced the utmost care to keep them comfortable, nourished, supported, and not too stressed.

Before transplanting, Susana explained that they would likely go through a bit of transplant shock, so we pinched off their little baby leaves—their seed leaves—the first ones that popped out of the embryos when the seeds germinated. Amputating those training wheels was doing them a big favor in energy efficiency so they could redirect their activity to adjust to their new surroundings and to the new growth to come. After guiding me through the process, she left me alone on the farm with ducks, chickens, geese, a dog, cats, and a multitude of wild and domestic plants. As I sat there and continued moving each small life into its new space, I felt a connection with each tomato plant. I was going through transplant shock, too.

Years ago, when I was researching organ transplants, I learned about rejection. If you get someone else's kidney or retina or lung, your body could say, "Thanks, but no thanks," and start its process of ridding itself of this foreign object. It seems like your body would accept what is good for it, but the body can have a mind of its own, and ultimately we have to abide by its rules.

Likewise, when a plant is crowded and needs a new pot or new patch of ground so it can spread out and be healthy, it's not an easy transition. Some just don't make it. Even with the right soil mix, water, and light, they might have gotten their roots bruised or crushed. Some will drop all of their leaves and later recover. Some just die.

For me, I chose to transplant myself into this new, vital, blossoming environment of Kentucky in the springtime. I didn't mind loosening up my roots, and I let go of a lot (maybe not all) of the old, dried-up stuff that no longer served me. And when it came to being alone on the farm, even supervising newer interns, my seed leaves/training wheels came off and I enjoyed that ride as well. But, for some reason, my body rejected my new living space. I managed the hay fever with natural remedies as I worked, and I was getting by all right. But when I woke up in the middle of the night, feeling my chest tightening down with the sensation of an asthma attack, it got scary. Instead of immersing myself in this moist, flowering forest full of unfamiliar pollen, I needed to be in a space where I could transition little by little to my new environment.

I chose to leave the beautiful Salamander Springs farm, the apprenticeship with the powerhouse Susana, and the vibrant community of Berea. It was painful to make that decision, to let someone else down, and to disappoint myself. Rejection doesn't just hurt when you are on the receiving end.

Fortunately, I have faith in things working out for the best. And that faith has led this country mouse to the big city of Lexington. I've transplanted myself again, and my tendrils of connections here are becoming fortified daily, with a healthy relationship and new friends reaching out and offering support. I'm being gentle with myself, allowing myself time to rest, heal, and enjoy my new space while I slowly gain ground in the field where I want to get myself established.

OUTER SPACE

Caring for ourselves—tuning into our bodies' needs, eating well, breathing, paying attention to our form while we work—can all be enhanced when we have supportive spaces to nurture our inner needs. Permaculture ethics equally balance care of people with care of the land and fair share. Integrating human-friendly spaces into our farms and gardens boosts our morale and welcomes others to enjoy them, if you invite them. You can always keep a special place just for yourself.

Remy says, "I think having a private, sacred space is especially important, whether you live in community or you see yourself living in the greater Community, with a capital C: the ecosystem community."

My small suburban backyard belongs to my partner, who is a natural landscape designer, so it is an oasis for small wildlife with a diversity of native plants. It is also a woodland escape from the urban atmosphere of asphalt, air pollution, and traffic noise that assault our senses on a daily basis. When we step out the back door, we inhale the oxygen-rich air produced by a massive pin oak tree and its understory of fruiting shrubs. A gazebo, a clubhouse/outdoor kitchen, and a fire pit give us spaces to meditate, cook food, work with plants, hang our laundry, and relax with friends.

However, my partner is a good eight inches taller than me, with, obviously, longer legs, so his cute stepping-stone arrangement to travel through the urban oasis is not spaced out for my steps. I either make awkward, longer strides to land on each stone, or I slide a bit off the edges, or I avoid it altogether by gingerly stepping on the earth between and beside the stones. You wouldn't think this is a big deal, but after spending a year passing back and forth along that path, I realized that it was not flowing well for me. I felt as if I were on a broken sidewalk when I preferred to think of the path as a wooded trail. I asked if we could replace the stones with something that would allow my feet to touch the earth, shoes or not, when I walk into the yard. I suggested wood chips because they

Do you have a special place outdoors where you feel comfortable and can relax? If yes, describe it.

If not, describe what your special place would look like. What are the obstacles in your way of having such an outdoor space?

would be easy to scrape off our shoes on wet days, and they break down to create a higher carbon content in the soil. Because he is a conscientious designer (and because he has to live with my complaints), he agreed.

We've replaced the stones and filled the path with mulch, and now I can glide through the garden without being preoccupied with my footfalls. I love the sound, smell, and springiness that this little change has brought to our pleasant space.

INVITATION TO SHARE YOUR SPACE

Farms do not have to be created for the ease of tractors. They do not have to be production-intensive outdoor food warehouses. Farms are living, breathing, interactive spaces, and we shouldn't forget that humans are part of them, too—humans like you and me, our families, our friends, and strangers who may be passing through. By extending an invitation to share a space that you have lovingly molded into a food-, fiber-, or medicine-producing farm, you can increase the value of your land—maybe not on the books, but in the hearts of those who visit a real working farm, whether rural or urban.

What is one little thing you can do to spruce up your garden or farm?

How do your pathways feel? Crowded or roomy? Bumpy or smooth?

Sketch your special space and the path you take to get there.

If you like sharing your space and you want to diversify your income by hosting, there are many ways to bring people to you. All kinds of agritourism ideas entice relaxation-seeking vacationers, day-trippers, or just mindful grocery buyers to your farm. Be creative about hosting them, and keep it simple and easy for you to manage.

I am drawn to the gathering spaces, outdoor kitchens, and fire pits of permaculture designs. In many cultures, there is less separation between work space and social space. The Hidatsa elder Buffalo Bird Woman was interviewed in the early 1900s, and she described the typical female farmer's role in her agrarian village. Amid their field of corn, they constructed a watcher's stage, which was a shaded platform where women could take a rest from the fieldwork, eat a meal, watch for garden predators (like magpies, gophers, horses, and boys), and sing songs to encourage the other women. Buffalo Bird Woman says, "We cared for our corn in those days as we would care for a child; for we Indian people loved our gardens, just as a mother loves her children, and we thought that our growing corn liked to hear us sing, just as children like to hear their mother sing to them." Imagine healthy young women gathering daily, singing lovely songs, and it's no surprise that soon young men would also show up. The maidens would make up songs that teased the boys with playful, often personal, lyrics, which they would accept with good humor.

Have you ever pitched a tent or strung a hammock up and spent the night with your favorite fruit or vegetables? If not, what are the obstacles in your way?

Invite wilder friends, too. In *Wildlife in Your Garden*, I share basic information about the who's who of wildlife with which we share garden space. You might be surprised to learn that certain animals have a bad reputation for no reason, and many actually benefit the ecology of a farm or garden. The invitation for wildlife comes in the form of habitat—food, shelter, water, and space—that meets at least some of their year-round needs. Every creature has an interdependent connection with the landscape and the air and water that flows through it. Remember that your farm is part of a much bigger system. A few ideas for providing happy human and wildlife cohabitation include:

- A simple bird blind, such as a camouflage tent
- A bench near a watering hole; the longer you sit still, the more comfortable wild birds will be with your presence
- A bat box on your home or barn
- Various birdhouses
- Large rocks in sunny spots
- Brush piles or hedges along the edges of your property
- Water features
- Bare earth for native ground-nesting bees

A pond can benefit more than livestock, especially if you rotate grazing and the animals are away from it for a while. Don't swim where livestock have been, but you can protect the edges of a stream or pond by not mowing, which will bring a multitude of beneficial insects, such as dragonflies; amphibians, such as spring peepers; mammals, such as bats; and birds, such as swallows.

SACRED SPACE

While sharing your land with colleagues, friends, and wildlife is a part of the joy of farming, having your own space is an opportunity to create symbolic meaning in your own sanctuary. *The Medicine Wheel Garden*, a book by E. Barrie Kavasch, brings together ancient wisdom and contemporary design to encourage gardeners to connect with their space in prayerful meditation. The medicine wheel gives a form for planting herbs that we need to heal ourselves and our community. It is generally a circle divided into four quadrants that represent the four seasons and four cardinal directions. There's much more to the significance of this design and its indigenous roots, and you have the freedom to assign whatever meaning you wish to the space. Your medicine wheel could be as small as a saucer with stones, flowers, seeds, and feathers to represent earth, air, fire and water. Or you can install a native herb garden full of perennials that bloom four different colors throughout the year. Make it as simple or elaborate as you wish. As you pay more attention to your own connections with your space, you may be amazed at the magical significance that the smallest things bring, and how you shape the space to fit you.

Medicine wheels originate in Native American culture and symbolize aspects of spirituality and healing.

DRAGONS AND DAMSELS

I had the pleasure of meeting an interpreter at a historic state park who shares the perspective of the Shawnee people during the late 1700s. She explained to me the significance of some of her silver jewelry. Around her neck hung a necklace with a pendant of a cross with an extra horizontal piece. On her finger was a simple ring with a stippled pattern that formed a grid. Both of these symbolized dragonflies, a powerful animal in the Shawnee culture and in other indigenous belief systems.

Dragonflies can move up, down, left, and right and can hover in midair, changing directions and speeds with great agility. Their multidimensional movements represent the passing between this physical world and the spiritual world. Not many beings can move easily between the two, so this creature brings a special power of transcendence. The Shawnee necklace with the double cross portrayed the double-winged dragonfly as well as the Cross of Lorraine as a way of incorporating Christian beliefs with their own traditions. The ring that the interpreter wore symbolized the path of a being who travels without inhibitions and goes beyond ordinary physical limitations.

Much more mythology surrounds the creatures classified in the order Odonata (the "toothed ones"). Just take their names, for example. The dragon and the damsel. Sounds like a fairy tale in the making, doesn't it? Once upon a time, three hundred million years ago, even before the time of dinosaurs, dragonflies and damselflies roamed the wild Earth (and that part is true!).

So, how can you tell a dragon from a damsel? Damselflies are not female dragonflies; the two are actually in different suborders within Odonata, displaying different behaviors and comprising hundreds of different species. For starters, dragons perch with wings outstretched, and most damsels keep their wings together (that's not meant as an innuendo). The dragon's eyes are close together; however, damsels have a distinct space between their reflective compound eyes. Damsels aren't in distress, but they are weaker aerialists compared to the dragons and their stealthy, quick maneuvers. They are petite, cautious, and more primitive in evolutionary terms.

Emerald damselfly

Dragonflies include the descriptively named families of darners, spiketails, emeralds, clubtails, skimmers, cruisers, and petaltails. Damselflies include families of narrow-winged, broad-winged, spreadwings, shadowtails, and threadtails.

If we compare it to a fairy tale, what goes on during the dragonflies' mating act can be a bit grim (pun intended). There's no Prince Charming when it comes to a "toothed one" sowing his seed. The dragon male can be a brute with his lady, grabbing her by the back of her head with his powerful anal appendages. If you see what biologists refer to as dragonflies "in tandem," with the tip of the male's abdomen at the base of the female's head, you could consider it a type of foreplay. He'll take her on a fly-about and then bend her so that the tip of her abdomen reaches his reproductive organs at the base of his thorax, making what is is referred to as a "mating wheel" that loosely resembles a heart, if you want to romanticize the act.

With 450 species of dragonflies and damselflies in North America, flying at speeds of up to 30 miles per hour, the male has to have incredibly keen eyesight to check out all the ladies and find a species match. Dragonflies (but not damselflies) are equipped with a lock-and-key system in their respective reproductive organs, so the wrong match is literally not a good fit. Often, several males will mate with one female, but the last one in is the first one out; the final sperm has the best chance of connecting with her egg. The male dragonflies compete not just to take a turn but to take the last turn with the female. They'll often guard her until she has deposited the fertilized egg. Sometimes, a male will even remove sperm from the female's ovipositor to give his own genes a better chance. His anatomy includes a barbed apparatus...well, let's just leave it at that.

Despite a wide variety of habitats and landscapes hosting them, dragons and damsels need to be near fresh water. Depending on the species, either a still, quiet pond or a babbling brook is essential. The female deposits a fertilized egg into water by dipping her abdomen down to the surface and releasing the egg. After hatching, the larvae, called ***naiads***, are excellent swimmers. Damsel naiads have dual-purpose gills that take in oxygen and act as fins to help them swim. Dragons have a more exciting mode of water travel—they propel themselves by forcefully squirting water out of the anus. That's quite a superpower.

Naiads resemble tiny sea monsters. They are powered by voracious appetites and fitted with a lower jaw like a steel trap that extends and retracts, jutting out and grabbing unsuspecting prey like the chompers in a "Hungry Hungry Hippos" game. They eat all kinds of small aquatic life, everything from mosquito larvae to minnows. Eventually, which can be years later, just like the ugly duckling, the naiad transforms into an elegant flying machine.

Dragonfly

Similar to amphibians, dragons and damsels are born in the water and crawl to shore when they metamorphose. They need the land as much as the water, where they cling to stems or bask in the sun on rocks during their teneral phase. If you want to encourage dragons and damsels in your pond, don't mow neatly up to the edge. Leave some standing native plants around the bank for them to hang onto.

At this point, they spend a few days becoming adults, and the ugly duckling/sea monster naiad emerges from its exoskeleton as a graceful creature with powerful, translucent wings. Now, they add the realm of the sky to their domain. Few birds can compare to these aerial predators. They eat all kinds of other flying insects, such as mosquitoes, wasps, butterflies, moths, bees, flies, and even other dragons and damsels.

It's plain to see why people who live closely with the natural world, such as the Shawnee, view these animals with respect and awe. Dragons and damsels perform equally well above as below, at the proper time of their life cycles. Dragonflies and damselflies are never harmful to humans, and they can be a great help in controlling other insect populations. And they pause, just long enough for us to glimpse their shimmering beauty. Insects may be annoying or frightening, and yet if we can move beyond fear and take a step toward appreciation, there is always more to their story.

YOUR PLANT ALLY

Like many Americans, I have some native blood as a fraction of my heritage, and my connection with that culture also feels like a small fraction, but it's one that I've been increasingly curious about. Finding one's indigenous roots and connecting with them can be an intimidating, humbling, and endless journey. For me, exploring native edible plants feels like the most respectful way to connect with the people who first learned how to survive on this land.

I attended a conference that brought much more depth to the stories around wild plants and our ancestors' relationship with them. At the Southeast Wise Women's Herbal Conference in North Carolina, I listened to elder Native Americans share their tribes' knowledge. Learning from tribal medicine women is a rare and special honor, yet Cherokee healer Dr. Jody Noé described the wisdom she shared as common knowledge for every household—or at least it used to be.

I also participated in a Native American herb walk with Cindi Quay of the Menominee Nation. Her perspective was simple and relevant in the midst of a busy conference full of information and experts. It provided an opportunity to become grounded in my relationship with healing herbs by focusing on just a few readily accessible plants in the vicinity. Cindi's most important piece of advice for anyone wanting to study herbal medicine is to become friends with one wild plant and spend an entire year (or more) really getting to know that one plant. It could be a flower that you have always loved the smell of, a weed that keeps popping up in your yard no matter how hard you try to get rid of it, or a shrub that suddenly grabs your attention on a quiet walk through the woods. No matter how you find each other, your plant ally will be your teacher if you allow it to be. Learn your plant's characteristics in every season and how it goes through the transitions. Learn about all its parts—roots, stem, leaves, flower, fruit, seed—and different ways of preparing or preserving each part. As you learn about each aspect, consider how it can help you or someone else.

On the walk with Cindi, she introduced us to a few plants that she had made friends with. I was surprised that some were invasive species, such as Japanese honeysuckle (*Lonicera japonica*). A little nectar from a honeysuckle flower can mean quick rehydration on a hot day. This brought back a fond memory from my childhood. A massive honeysuckle vine grew outside our kitchen window on a trellis, providing shade from the southern sun throughout the hot, dry Texas summers. I would pick the little flowers, suck out the nectar, and feed the blossoms to my dog. It turns out that parts of honeysuckle hold potent antibiotic and anti-inflammatory qualities. Soothing, indeed.

Cindi reminded us that these exotic and opportunistic plants were originally introduced because of their healing and nutritional properties. Despite the fact that they indicate and exacerbate ecological imbalance, we have the choice to accept what they have to offer and benefit from their abundance. Dr. Noé's Cherokee elders taught her that herb harvesters follow the "rule of four." They may harvest only after seeing a particular type of plant four times. The fifth time, they make an offering and ask the plant for permission to take some of it. This way, they remove only one-fifth or less from a local population. At least with invasive species, we know there is plenty to harvest, so it's no problem following the rule of four.

Back at Dr. Noé's tent, the special plant ally she shared with all who attended was spicebush (*Lindera sp.*). A tea made from spicebush leaves and twigs is said to make friends out of enemies. She told us that it indicates that fresh water is somewhere nearby, which is a precious resource for all people. If you make spicebush tea for yourself, someone you don't really like, or someone you love, Dr. Noé says, "It opens your mind, and it opens your sinuses."

Dr. Noé and Cindi Quay helped me open up to the idea that regardless of my heritage, plants themselves are my teachers. Watching animals interact with them, being aware of the subtle nuances that draw me closer, observing more deeply, sensing what I might be needing or seeking when a certain plant catches my attention—these intuitive cues helped me meet my plant ally for the coming year: a pitch pine tree. When I explained my discovery to a friend, I said that I had been feeling "sapped" of energy after a long day at the conference when I tripped on a root. I looked up, and there was a tall, protective-looking tree with sticky sap, the healing aroma of which revived me as I rested in its shade. She laughed at the unintentional pun ("sapped") that I hadn't even noticed!

Dr. Noé said to the packed conference tent, "All of you are indigenous to someone, somewhere." In reality, your "tribe" is who you feel most connected with and influenced by. The cycles of the year bring harvests, celebrations of the bounty, and surplus to share by serving loved ones and strangers alike. Brew some spicebush tea, take a deep breath, and drink to your health.

Pitch pine needles and pinecones.

Harvest and Share the Bounty

When we are blessed by a surplus, we can give freely. An attitude of gratitude for abundance feeds back and continues to support our needs. Our times of harvest are great reasons to gather help and celebrate all that nature and our hard work have provided. When we pull fruit from a vine, we've good reason to rejoice at the culmination of watchful anticipation and prayers for fertility. The harvest results from the intersection of perfectly timed rain, wind, sunshine, and temperatures, along with protection from harm. It's a wonder we don't shout "Hallelujah!" with our mouths full at every bite of food! Whether we praise the functional mechanics of science or attribute our food to the miraculous life force from the divine, we can all agree that harvests are worth celebrating.

While we think of the fall as harvest season, the harvest happens all year 'round. From the spring's first microgreens to the summer's first handfuls of strawberries and snap peas to the cucumbers that relieve the heat with their cool crunch to the golden warmth of the fall's first butternut squash—give in to the urge to cheer each change and welcome each new fruit and vegetable as if it's exactly what you hoped for. Congratulations! It's alive!

On the flip side, in November, when I have been harvesting tomatoes since July, there can be too much of a good thing. I've put by pasta sauce, ketchup, and salsa and packed the freezer with diced tomatoes. During a prolific tomato season, the ten 50-foot rows growing at the farm where I work part-time taunt me every time I pass them. My coworkers and I averaged 100 pounds of tomatoes every week for two months! I'm no longer excited about pulling cherry tomatoes high and low, hidden among itchy little leaves.

When I come home from work, there's another pile of tomatoes waiting for me, which my partner brought home from the garden at his office. I hate for any food to go to waste, and there's nothing worse than the smell of a rotten tomato (think dirty, sweaty socks). In short, the joy of celebrating tomato harvests has long passed me by. However, I will be happy to pull out a bag of frozen diced tomatoes in the middle of the winter, when it's time to make a big pot of chili and cornbread. Yes, then I will celebrate and be glad we didn't let them rot on the vine.

COMMUNITY SUPPORTED AGRICULTURE (CSA)

The CSA concept is more than just the basket of produce that loyal customers get once a week. There are many variations on the idea and, in reality, it's mutually beneficial: the farm supports the community, the farm is part of the community, and the community supports the farm. There are many ways to relieve yourself of the heavy lifting during a harvest. Following are a few ideas.

Harvest parties: Harvesting can really bring together a community and help a farmer get a lot of work done in a short time. Make it a potluck affair and consider providing camping spots, if possible, so your guests can pitch in for a good day's work and then some. Establish an annual harvesting event that you will all look forward to.

Work share: One farm near me hosts Many Hands Mondays, which draws in CSA members, friends, and anyone else who wants to get a feel for farm work one day each week. Other CSA models include work-share agreements that include a trade of labor in exchange for a reduced cost on the share basket.

Pick and prep/farm to table: Customers can pay to learn about where their food comes from, connect with it in the field, and harvest it. With a chef as a guide, guests can then indulge in a cooking lesson that culminates in the freshest meal they've ever eaten.

CSAs allow customers to be part of the growing and harvesting process.

U-pick: This might be the oldest trick in the book. Sit back and let your customers do all the work for you and then pay you for the privilege of getting a taste of farm life. During the busy season, engaging your community for labor does require planning and coordination, but it can pay off in a big way by lightening your actual physical workload. Working with others will also strengthen the neighborhood, forming the community bonds that a healthy farm needs.

Lisa Munniksma reflects on her dozens of farm experiences as a traveling freelance farmer. "Even if you have strong family support, it's also really a community thing. I've witnessed a lot of really strong coming-together." Many times, a matriarch of a farm family is the one to coordinate work parties. Lisa says, "It's not just a woman thing, [but] it's largely driven by women. I think that it's really inspiring to see other women active in their communities. That's something that I see a lot, traveling and working on farms."

Here are some more considerations about sharing the bounty while sharing the work.

- Working together can teach the next generation the skills they will need (without making it seem like a real lesson). Decide how you will include children in the group and be sure that they have adequate supervision.
- The end-of-year harvest time is a perfect time to advertise next year's CSA and encourage members to re-enroll.
- Save up some of your most menial chores, those that make your body revolt at the thought of doing, and let the many hands make light work of it.
- Try to group skilled and unskilled workers in beneficial combinations for the jobs at hand. "Each one can teach one" is a good policy for managing a diverse set of helping hands, allowing them to build relationships among themselves and relieve you of some of the management.
- Remember that volunteers are benevolent, but they also have needs. Provide breaks, restrooms, and refreshments to keep them comfortable. Having extra gloves on hand is always a good idea.
- Make it a tradition. Continue to nurture the connections you make and stay in touch with updates about the farm throughout the year so that they stay interested and return to your CSA.

WASTE NOT

Some of the best food often consists of what some would consider "throwaway" items. Turnip greens, pork cracklings, bone broth—these rich ingredients satisfy the soul and deliver powerful nutrients to our cells, so don't undervalue parts that you might otherwise toss aside.

What memories do you have of seasonal harvests? What foods do you associate with harvests?

What personal celebrations and seasons mean the most to you?

What have you harvested this year? Do you have records of the amounts planted or animals raised? If so, list them here.

An overabundance of food does not necessarily mean waste. Whether you are tossing rotten fruit to your pigs or setting aside produce that didn't sell at market for a charity, be mindful of the marginal.

A "food desert" is a residential area where a grocery store within a safe walking distance is more than one mile away. In a perfect world, every convenience store and pharmacy would sell fresh, affordable, local food. Even then, access to these markets would still be a challenge for many people. Imagine riding public transportation with two small children, carrying enough groceries for your family for a week, dodging traffic, and walking a mile home. In the rain. In the dark. In the cold.

Food pantries work wonders in supplying food to those who need it, and donations are essential to these institutions. Further, donations of fresh, wholesome, local, organic food is a great need. Glean KY is a nonprofit organization that bridges the gap between farms with too much food and inner-city eaters with too little nutrition. Their small

army of volunteers gathers donations at the end of the day from farmers' markets, from farms themselves, and from grocers. They collect fresh produce and take it to food banks as well as to community leaders. For example, a volunteer allows her home to be the distribution point for families in her neighborhood who can use the extra fresh food at no cost.

In a commercial sense, food hubs are filling this need as well, delivering produce from farms to buyers. A wide variety of models exist for food hubs, some of which are essentially CSA delivery services that bring farm food to the inner city, while others aggregate larger quantities from multiple small farms to enable purchasing for wholesale buyers.

Aunt Judy chimes in with her wisdom: "Invariably, you'll have things that fail, things that succeed, and things that 'oversucceed.' How do you deal with that? What do you do with an overabundance? Learn who to share with."

What of this am I going to let go of? Should I let go of it? If I hoard it, what's going to happen?

What resources exist in your community for sharing the bounty? To whom do you donate excess food?

SHARING SEEDS

What is a seed? My college botany professor taught us that a seed is "a baby with a lunch bucket." It carries a complete starter kit for life in the world—a protective coating (testa), a foot to get rooted (radicle), an arm or two to reach out to the sun (cotyledon), the beginning of a stem for structure (hypocotyl), a flower bud for creation of future plants (plumule), and a supply of starchy energy to fuel the early growth spurt (endosperm).

Seeds hold metaphorical, ancestral, and physical energy. These tiny nuggets of life that tumble out of ripened fruit contain the genetic code for the cucumber, zinnia, bamboo, or carrot that will germinate, sprout up, mature, and offer its blossom to pollinators for fertilization. No wonder seeds have been held in such high regard by civilizations throughout time. They continue to inspire poetry and spiritual practices, and they bind us in a deep relationship with plant reproduction for the good of our shared survival.

In the wild, seeds are spread by birds, mammals, wind, and water. Seeds are relocated to new gene pools, and they strengthen diversity. When human cultures began domesticating plants, we chose to narrow the infinite number of possible variations and keep the seeds of the plants with characteristics that we liked.

Harvest time is also seed-saving time. Seed saving is a practice of editing as we choose what is worth keeping: plants that have the strongest, most resilient characteristics in the face of adversity and that produce the tastiest, most pleasing food. Seed saving used to be a common practice before packaged seeds became widely available. Shaker communities in the nineteenth century were some of the first agriculturalists to meticulously keep track of their seeds, package them with planting instructions, sell them to the public, and provide guarantees on their products.

Seed banks and seed libraries offer communal spaces to keep good seeds safe and share the bounty. Heirloom varieties are gaining favor with the general public for their interesting aesthetics and flavors, and with farmers for the simple wisdom of the right plant in the right place. Heirlooms, as the name implies, have been selected over many generations for the qualities that ensure their survival in the location where they are planted.

Seed swapping is making a comeback. In this ancient practice that is surrounded by ceremony in native traditions, the seed is a sacred symbol as well as a source of life. At a Food and Justice conference I attended in Taos, New Mexico, presenters shared a seed ceremony with attendees. Prayers of gratitude, surrender, hope, and protection washed through the circle of farmers surrounding special vessels that contained the year's best seeds. The blessings included a frank and educated discussion of seed sovereignty.

The food and seed sovereignty movement is worldwide. Vendana Shiva and her organization, Navdanya, are a "women-centered movement for the protection of biological and cultural diversity," as stated on the website, www.navdanya.org. "Conserving seed is conserving biodiversity, conserving knowledge of the seed and its utilization, conserving culture, conserving sustainability."

The seeds we plant hold all the genetic code for the coming generations: nourishment, belonging, heritage, food, freedom. All of the genetic information passed on from previous generations and all of the genetic information that our successive generations will inherit is in our seeds—whether we hold those seeds in our wombs for our ancestors and to pass on to our offspring, or in the fruits of our foods, or in the words we speak and the attitudes we spread. Women are the holders of great power.

When saving seeds, every type of crop is treated a little differently. The following are some general guidelines to get you started.

- Leave the best fruit (however you define that for your needs) on the plant until it is fully mature. For cucumbers, tomatoes, and peppers, look for their colors to indicate that they are beyond ripe. For example, cucumbers or peppers that are intended to be harvested when green should pass into their red or orange phases. Similarly, leave eggplants on the stem until they turn golden yellow.

Do you have any heirloom seeds that you are saving? If so, list them here.

What do they mean to you?

If you were forced from your home and could take only one seed with you, what would it be?

What is one seed, literal or metaphorical, that you want to save from the past season or year?

What will be the seeds that you sow next season or year?

- Lettuce and similar plants are in the composite family; their flower discs hold hundreds of seeds that should be harvested just before they fall out on their own.
- Legumes and pulses do best if they dry to the point of brittleness on the vine.
- Brassicas also have pods full of seeds, but they should be harvested when the majority of pods are brown and soft, not brittle.
- Tomatoes, pumpkins, and squashes have a coating on the seeds that inhibits germination. Soak them in water until the seeds sink and the coatings float and then dry the seeds thoroughly for about two weeks.

Self-pollinated plants are the easiest for beginner seed savers to have success with; these include the plants in the foregoing list except for squashes. If you have two different types of squashes or pumpkins in the same field, you could end up with a hybrid, and what grows from the seeds will be unpredictable.

From the harvest-season corn and squash to the New Year's black-eyed peas that bring good luck, there's a deep history associated with the food we grow that must not be forgotten.

SEEDS OF SLAVERY

The seeds we sow in our gardens have a deep history that isn't necessarily a pleasant one. Imagine being stolen from your home and forced to leave behind every physical scrap of connection to your land, your people, and your family. What could you possibly take with you that wouldn't be discovered in invasive, cruel inspections? Perhaps a few seeds—tiny little kernels of life and hope.

Millions of slaves were brought to North America, and some managed to bring a few precious seeds from home, nestled into the coils of their hair or tied into necklaces or bracelets. Think about life without foods like okra, yams, licorice, black-eyed peas, sesame, and watermelon. These staples of our modern diets are part of our culture now because slaves smuggled them and cultivated them, sometimes in secret.

Angelique "Sobande" Moss-Greer, herbalist, certified holistic nutritionist, and great-granddaughter of a slave, says that many slave owners would not allow slaves to have their own gardens. But many whites were afraid to go into dense forests, so that is where slaves could find a refuge from their masters and sow their own seeds among the wild plants. In concealed woodlands, slaves grew heirloom foods that had traveled with them from across the globe. These foods would nourish them, remind them of their heritage, and create a lasting legacy in their new communities, which were made up of dislocated people from dozens of distinct ethnic groups. Angelique's great-grandmother barely survived aboard a slave ship that carried her pregnant mother. She began the harrowing journey as a fetus in her mother's womb and arrived as a newborn.

So many humans endured agony during the transatlantic slave trade, particularly this baby, who was born in transition from freedom to slavery; this, in Angelique's perspective, created a type of generational trauma. The scars live on, transmitted to descendants on a cellular level, and need deep healing. She is taking what her elders taught her about herbal medicine and traditional foodways and using it to heal others.

The slave seeds were hidden treasures holding a small but precious power. In your season of seed-saving, in the pause before the season of seed-starting, take a moment to reflect on their heritage and significance.

IS THERE A MARKET FOR UGLY FOOD?

It's Sunday afternoon. I'm in a spacious, elegant kitchen with an eclectic group of new friends, armed with knives and chopping boards. We're digging through boxes of green and purple cabbage leaves, an armload of mangoes with the occasional bruise, and bushels of organic apples. Conversations bubble up around how to process this abundance of fresh food and make meals for the week ahead, how to make haste with waste. (How much salt do you add to the chopped cabbage? How many teaspoons are in a tablespoon? Does anyone have cilantro for the salsa?)

All this food was thrown out at a local market. My resourceful friends who manage community composting made an arrangement with the market, and with a Facebook post and access to a fine kitchen, several of us converged to salvage and share the bounty. The nine of us will eat Dumpster-dived meals this week and thoroughly enjoy the mango salsa, sauerkraut, and baked apples. Back at home, my breakfasts are ready as well. The freezer is full of smoothies I made from produce on the discount shelf: organic bananas for 38 cents per pound and a bag of oranges for a dollar. I am all about ugly food.

The homogenous scene of shiny, perfect, and consistent fruits and vegetables that greets you in a chain supermarket betrays the truth about food waste. Globally, about a third of all food grown is never consumed. In America, estimates on produce waste push the 50 percent mark. The cracks that good food falls through include industry standards for grading, spoilage on the long road from farm to table, and the mechanization of harvesting. Approximately 96 billion pounds of food per year go to waste in America, with much of it ending up in landfills, and we pay $1 billion per year to dispose of it.

The good news is that, globally, consciousness about appreciation of fresh food, perfect or not, is rising. Major efforts to reduce waste in food systems have been spreading throughout Europe, notably with the support of the European Commission's declaration of 2014 as the "European Year against Food Waste." Going back to 1999, the US Environmental Protection Agency (EPA) and USDA published a joint report, titled *Waste Not, Want Not,* that addressed the major discrepancies in this country. They constructed what they refer to as a "hierarchy of food recovery and waste redirection," which prioritizes solutions:

- Recovering food to feed hungry people
- Providing food to livestock farmers or zoos
- Recycling food for industrial purposes
- Composting food to improve soil fertility

Noble as these goals may be, it's tough for people to get excited about policies. The fun comes in the marketing. We have to tweak public perceptions about food safety, and we can use advertising for a greater good. Canada's largest retail grocery store chain, Loblaws, launched a campaign to

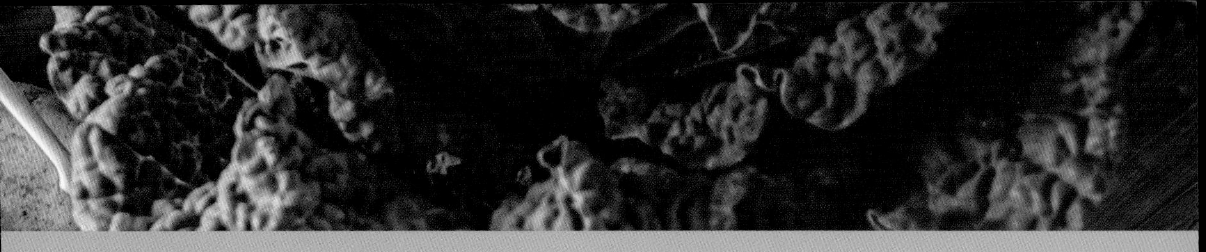

market their ugly apples and potatoes by bagging and branding them as "No Name Naturally Imperfect." The French grocer Intermarché called its similar campaign "Inglorious Fruits and Vegetables," and it proved to be a success, attracting 24 percent more traffic to stores. On the wholesale end, products such as vegetable soups, fruit juices, and peeled and diced winter squash open up value-added markets that aesthetically challenged crops can easily fill.

As a farm volunteer one summer, I helped harvest a field of sweet potatoes. With surprise and disappointment, we couldn't believe that every single tuber we pulled out had been nibbled by rodents! It looked like a total loss. However, the community-conscious farmer knew of a social service agency that trains and employs disabled people in processing food. We took them our pallets full of sweet potatoes, and they peeled and packaged them for local restaurants to make sweet potato fries.

If we had taken those potatoes to market, we might have been able to sell them, and we would have ended up donating them as a last resort. Farmers' markets are less interested in restricting food appearance than making food accessible. Many work in tandem with emergency food shelters. You can find bargains at the end of the day, when collard greens are wilting but will perk up with a quick dunk in ice water. A basket full of pockmarked summer squash labeled "seconds" can be a profitable deal for both the farmer and the shopper.

Second Harvest, the national food bank network, estimated that 21 million Americans depend on food donations, but charities often run out of fresh food to give them. Gleaning organizations can help connect the produce with the people. The aforementioned nonprofit called Glean KY (formerly Faith Feeds) picks up what gets left behind at farmers' markets, at grocery stores, and even straight from the farm. Volunteers deliver the food to feeding organizations and shelters, bridging the gap between the one in six Kentucky residents who are at risk of hunger and the supply of edible and nutritious food that is at risk of rotting.

If you want to make the most of the food our earth provides and do something to help reduce food waste, you have many options. If you are a consumer, buy the seconds and uglies to show the retailers and farmers that you accept that beauty is only skin-deep. If you are a grower, try to negotiate with your markets, whether wholesale, retail, or direct-to-consumer, and try to be flexible on pricing. If you are the rest of America, quit being snobs and embrace fruits and veggies with bumps, bruises, and twisty parts.

HANDS-OFF GARDENING

Thinking about fall cleanup? For the most part, you can just wait and do it in the spring. This fall, give yourself permission to neglect your garden just a little. Uncomfortable with that idea? Read on to understand how allowing nature to run its course can be the best thing you do this time of year.

Sherry Maddock lived in Lexington, Kentucky, and ran 4th Street Farm. It is a rich and diverse urban farm on a tiny downtown lot where an abandoned house once stood. She mulched, planted, weeded, and cared for this space for the better part of a decade, turning it into a food-producing oasis. With heartfelt generosity, Sherry freely shared the abundance of greens, beans, fruits, and herbs by the bagful. However, Sherry practiced loosening the reins on the community garden when she and her family began preparing to move to Australia.

Broccoli, cauliflower, kale, and cabbage are popular plants in the hardy genus *Brassica*.

During the winter and early spring of that year, she stayed out of touch with her garden for several months. When she returned, the unintentional hands-off experience showed her what can happen without human intervention. "In the fall, I had spectacular greens, just a huge array, a very diverse array of [*Brassica*] greens. Every bit of that has come back on its own; either they self-seeded, or maybe some of the seeds that didn't germinate came up later. It was lovely to have a self-sown garden this spring! I didn't have time to get out there and plant things. So the garden provided for me."

Sherry now believes that a garden consists of a two-part relationship: stewardship, or deeply involving herself in the environment and shaping the space; and "unattachment," giving the space and individual plants the opportunity to flourish on their own. "I may only want kale because I want those leafy greens in the spring, and when that's finished, those plants are often discarded. Well, it's beautiful to watch kale finish its cycle of life as it grows long stems and then flowers and then forms seed heads. It's doing what it should do, and it's good that we've left it alone, because we're not the only ones who need to benefit from what comes to the garden. We may be finished with the kale leaves, but the pollinators love those beautiful yellow and white flowers from all the Brassicas." Sherry's observations highlight some of the great reasons to go hands-off in the fall. It benefits the wildlife, the soil, and the plants themselves.

Wildlife: The timing of natural cycles coincides with our own physical needs and energy levels. At the end of the growing season, our bodies are tired and the days are growing shorter, cueing us to move toward rest. In the springtime, with increasing light and warmth, we are just itching to get outside, soak up some vitamin D, and wake up our stagnant muscles. In general, spring is the natural time to clear out plant debris to make way for new growth.

Birds and small seed-eating rodents are also designed to make the most of the fall harvest. If left to natural processes, these creatures will do the plant gene pools a big favor by dispersing the seeds far and wide as they fly above or tunnel below ground. Most birds are not year-round seed-eaters; they eat soft grubs and caterpillars in the spring (also excellent timing). So, if you are cutting down your sunflowers and perennials while purchasing bags of seed to stock your bird feeders for the winter, you may want to rethink your system.

Prevent your garden from becoming a "food desert" to native wildlife. Over millennia, dynamic and interdependent native insects have coevolved with native plants. As a chemical-free gardener, you can make a big impact on the overall health of these insect populations, the large majority of which are considered beneficial for vegetable, fruit, nut, and herb production. You can intentionally provide food for insects in the form of native plants, which will in turn attract more beneficial insects, which will control unwanted ones.

If you provide habitat for native pollinators, you will be rewarded with fruits (and vegetables) of your labor. Two types of insects to focus on hosting are native bees and parasitoid wasps.

As much as 85 percent of all crops require a pollinator to work its reproductive magic so that fruits

and vegetables form. Collectively, native bees make up a powerful pollinator force, although they are overlooked in the shadow of domesticated honeybees. Approximately 4,000 species of bee are native to North America, spending much of their lives in solitary cells underground, emerging to forage in the spring and summer. When they do visit flowers, they are more effective pollinators than honeybees because they are less meticulous about their work and end up spreading more pollen to more flowers. They are also better suited to the climates where they are working, as are most native organisms.

Regarding wasps, if your tomatoes were devastated by hornworm caterpillars, don't spray them. The enemy of your enemy is your friend. Braconid wasps, for example, meander about, lovely and harmless, as they sip nectar from your carrot-family flowers. But just wait until they unleash their secret weapons. These wasps will insert eggs into live hornworms or other soft-bodied targets. When the eggs hatch, the tiny

(continued on page 168)

larvae will suck the life out of the caterpillars, slowly devouring them.

Attract parasitoid wasps to the garden by planting herb and flower species that supply nectar and pollen, including Queen Anne's lace, dill, cilantro, and fennel—all plants you may be tempted to cut down at the end of the season. There's good reason to let them stand and attract these beneficial, stingless wasps—they will also take care of aphids, squash bugs, and stink bugs, among other garden pests.

Many other types of beneficial insects lay their eggs in the fall under the shelter of perennial foliage. Don't dispose of your trimmed plants (unless they are weeds, diseased, or pest-ridden). You could be destroying nurseries that protect beneficial insects' eggs. Instead, try creating a border around garden beds with your excess stalks, mulch, straw, and leaves.

Providing cover can be as important as planting food plants for insects. Many insects go into a type of hibernation called *diapause* in which they slow down all body functions as they wait out the winter. To do this safely, they find protection in hollow plant stems, in logs, under rocks, and under leaf litter. Even cutting the stalks and piling them in a corner is better than getting rid of them.

Despite the importance of protecting soil, also try to allow some bare ground along outer edges. Ground-nesting bees will use these areas, which are less likely to be disturbed by gardening. Provide nesting areas for mason bees and carpenter bees in hollow stems or logs, and they may leave your garden shed alone.

The following great plants will attract, feed, and shelter beneficial insects year-round:

- Goldenrod
- Coreopsis
- Black-eyed Susan
- Yarrow
- Purple coneflower
- Sunflower
- Salvia
- Penstemon
- Bunch grasses
- Echinacea
- Culinary herbs, such as lavender, sage, basil, and oregano

Black-eyed Susan

To find specific plants that will host your ecoregion's pollinators, use the zip code search function at the Pollinator Partnership's website, www.pollinator.org/guides.htm.

Sherry attests to the impact of providing great insect habitats: "People marvel when they come to our space in the summer and see vegetables growing beautifully, and they say, 'But you don't use any pesticides,' and I say, 'Absolutely not.' Our beneficial insects, they're all taking care of everything for me."

If you don't like the thought of more bees and wasps, remember that insects are interwoven throughout the entire food web, and their health and abundance affects every other creature as well as your produce.

Soil: Initially, building healthy soil requires hands-on work. Layering grass clippings, manure, shredded leaves, newspaper, cardboard, food waste, compost—whatever you put into your mixture, it takes some sweat equity to build this foundation. Putting time into that work will pay

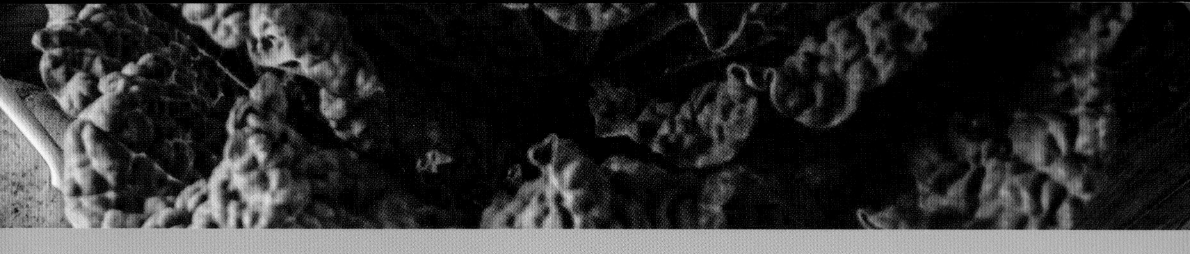

off well. Leaving plants standing in the autumn and planting cover crops can protect your investment.

To block prevailing winds in the cold months, either plant a hedgerow out of perennials or leave annuals standing in place. Their roots anchor the soil, and the vertical stalks and dried leaves buffer the harsh wind. Leaf mulch won't blow away, and you will appreciate the windbreak that slows down blustery breezes during fall harvesting.

Sherry recommends letting perennials, such as asparagus, fill this niche. "A lot of people will clear it in the fall, but I leave those ferns because I think it's a good windbreak. For the soil, it's more protective to have that dry, brittle fern left. In the spring, I clear it when new growth is ready to come up." Protecting the soil with these natural umbrellas also helps beneficial insects whose eggs or chrysalids are overwintering in the ground.

Hands-off also means feet-off. Remember not to tread on your beds, even when it looks like nothing is growing, Stepping gently and staying off wet soil will prevent the spread of disease and minimize soil compaction. Imagine wetness as a "do not disturb" sign on the ground.

Mulched paths between beds provide the access you need without disturbing the soil. Fall-planted cover crops are considered green manure because of the work they do to improve soil fertility. *Chop and drop* is the term for planting cover crops that can be harvested and used as either pathway mulch or turned into the soil as fertilizer. They provide certain benefits for the garden; for example, Siberian pea fixes nitrogen, and comfrey delivers the trace minerals it transports with its deep taproot. If you are looking for some good ol' biomass, then winter wheat and different types of ryegrass might be worth trying.

In the spring, be prepared to chop and drop the crops before they set seed, or you could wind up with too much of a good thing. Wait a couple of days until they turn brown and then work them into the soil. Wait another few weeks before planting so that decomposing greens won't tie up nitrogen in the soil.

Local extension offices can recommend the best selection of cover crops for your site and when to plant, which should be at least four weeks before a killing frost. Don't shy away from planting cover crops that won't tolerate a hard winter freeze; they'll simply do you the favor of dying off before spring arrives and without going to seed. With the use of cover crops, this hands-off approach allows autumn-sown seeds to help you prepare beds for the spring by revitalizing the soil.

Plants: The only thing better than hands-off gardening is being able to harvest and eat fresh veggies all year-round. Some cool-season crops can survive through the winter with a little help. Here is a list of some common crops that you can allow to bolt (quickly produce flowers and seeds) and self-seed and then enjoy their tender greens early next year (ask your local extension office about the specific varieties best suited for your area):

- Mâche (salad green)
- Arugula
- Lettuce
- Kale
- Endive
- Radicchio

(continued on page 170

- Cabbage
- Broccoli
- Legumes

Don't give up on your little baby plants that overwinter. Expect to see little to no growth until the soil warms up. They will conserve their energy and show only a few true leaves, which also reduces their risk of frost damage. They aren't exactly growing during the winter anyway, but they are putting energy into their root systems. Many vegetables can tolerate moderate frosts and a little snow, but be prepared to cover the tender youngsters if a harsh storm is predicted.

There's no need to dig up all your root vegetables and then store them indoors when you can use nature's root cellar by leaving them in the ground. Sugar levels rise when the temperatures drop, which results in sweeter carrots, beets, rutabagas, and parsnips. To help them make it through the cold fall and winter nights, cover any exposed shoulders with soil, but let the green tops keep their heads exposed to sunlight.

If temperatures are seriously dropping or you live in a more northern climate, cover the crops with a deep layer of mulch or blankets at night and remove them during the day.

Cool-season greens like spinach and chard can stay in place for winter salads, especially if you create a mini-greenhouse out of straw bales and window panes or cover them with hoops and clear plastic. When you plan your garden, situate these plants in the sunniest spot and near a wall or fence to act as a heat sink.

To be clear, you should not leave every plant unchecked in your garden this fall. Cut the seed heads off of weeds and get them completely out of your garden. Also, clean up areas where any major insect damage or diseases really hurt your crops. Pull out plants completely and burn them or dispose of them in your municipal yard waste or household garbage bins.

Sherry has learned that she can't allow even the most-loved plants to run amok at 4th Street Farm. Among the wilder plants she chooses to control include the ubiquitous mint, garlic chives, the cup plant (*Silphium perfoliatum*), and a native goldenrod.

She compromises by keeping fennel for the butterflies. "I have it, and I allow it. It has spread, and it's hard to get out because of its root nature. I tolerate it, but it kind of makes me irritable because it crowds out my herb bed, solely because of the black swallowtail [butterfly]. This is like cohosting with your garden. If you long to host these butterflies and watch them emerge, then you have to provide their habitat, and they love fennel and dill and that whole family. So I've allowed the fennel."

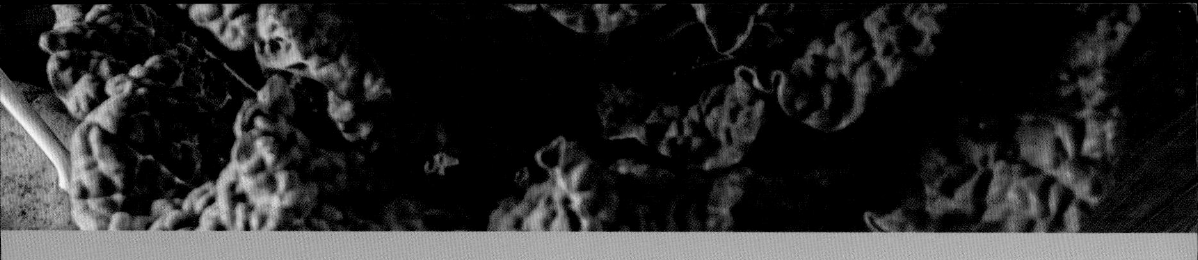

There are good reasons for letting some plants bolt and go to seed, one of which is seed-saving. This practice of handing down heirloom seeds that were bred in the field, not in a laboratory, used to be common. Breeding in the field simply means that you select the best plants from the garden plots each year. Over successive generations of doing this, you wind up with seeds that are bred not just for your area but specifically for your site and the microclimates unique to your space.

You don't need any special tools or professional horticulture education to save seeds and breed your own specialized varieties. It does, however, require a little technical expertise, and you can learn by joining up with local seed-saving groups and referring to a guidebook, such as *The Manual of Seed Saving* by Andrea Heistinger. The beauty of this type of plant breeding is that nature does the work for you. Andrea summarizes, "Whatever the time span, the environment does the selecting through variables such as length of growing season; type and frequency of precipitation, drought, and cold periods; and intensity of light, to name but a few."

The result of allowing nature to run its course, to some extent, in your fields results in even less work needed as generations of heirloom varieties improve upon the previous versions. They continue to adapt genes to thrive where they are planted, ultimately needing less fertilizer, pest management, or special care.

In my opinion, one of the best aspects of seed saving is that it requires keen observation skills, which are refined along with the varieties of plants. Which individuals were least affected by a disease? Which ones bloomed first and longest? Which ones produced the juiciest fruits? You are the judge.

In a video on his website (www.geofflawtononline.com) about how to judge a garden's success, permaculture consultant Geoff Lawton points out that there are a multitude of methods to garden organically and sustainably, and all of them work. The key is to keep building the soil, and you measure success by the amount of work you don't have to do. "Over time, overall production becomes higher in quality and more diverse," Geoff says. The garden improves by your interaction, and your effort diminishes. Sherry's reflections echo this idea. She has seen an increase in her garden's perennials over the past few years, enough to divide and share with neighbors and friends.

Sometimes the hardest thing for gardeners to recognize is that we're not really in charge. We work in collaboration with nature, making it easier for seeds to grow and filling our harvest baskets as a reward for enabling nature's habits.

As Sherry gazed out her kitchen window at sprawling carpets of thyme humming with pollinators, she shared her thoughts on leaving this space. "I can't hide from the things that hurt, so I celebrate them. Last night, we sat outside, and I looked at Asian pears and green figs, and I said, 'You know, we're watching fruit grow that we will not eat. We won't see it ripen.' And there's something beautiful about belonging to something much bigger, [where] the end point isn't 'I will eat that.' We will contribute to the life of the next family and their friends who will walk into the garden."

Chapter 5

Good Teacher, Good Student

We celebrate our abundance of food and the fertile lands that provide it, yet it is harder to bring the less tangible fruits of our labor to the forefront. Our knowledge is priceless. The wisdom from generations before us, the simple insights we learn from children, the colleagues who bear witness to our personal growth and support us through our development—these are worth celebrating as well. We may acknowledge them in quiet ways, which is appropriate for the slow, steady methods that bring us these gifts. Nonetheless, to be a leader, a mentor, a guide, a teacher is a great honor that must be earned. To be a student is a lifelong endeavor and is how we should all approach our stewardship of the earth.

We're past the days when society expected an educated female to become a teacher or a nurse. Ironically, farmers are both teachers and healers. Women farmers often identify themselves at some point as one or the other, and probably both.

I have been impressed when hearing presentations by indigenous elders and women who are leaders in natural healing. They begin identifying their lineage, and while this sometimes takes up a good portion of the allotted time for their talks, it conveys the honor that their teachers have earned. They list the people who have influenced the path they are on and remind us that we are all connected. The knowledge they share with their audiences comes from many generations before us, and we who listen will affect the coming generations.

Angela Wartes-Kahl and her husband/farm comanager, Garth, are leaders in organic certification and apprenticeship programs. Angela and Garth telecommute from their rural Oregon farm as organic inspectors and also have an organic consulting business. Angela describes her work, saying, "We help large processors start the organic certification process from the beginning and develop products." In addition, she is the fiber and textiles coordinator for Oregon Tilth and helps textile mills get organic certification. Their Common Treasury Farm, which Garth and another partner founded, has been certified since 1993.

How have you shared your knowledge? Who are your current students?

Who have been your past mentors in farming?

Who are your current mentors in farming?

List the leaders in your immediate community with whom you have a personal relationship.

List the authors, speakers, or role models you look up to.

The apprenticeship program that Angela manages is intentionally created to train, as she says, "not just more farmers, but more organic-certified farmers." She believes that this is more than a job. "Our dedication to organic is lifelong and incorporates all aspects." She leads new farmers in applying the practices and living the lifestyle.

Unlike many training opportunities on farms, Common Treasury Farm emphasizes the extracurricular activities. Angela says, "Apprenticeships limited to free labor didn't realize their full potential and created a need to expand and understand the business." Angela and Garth make an effort to facilitate networking situations for young farmers by helping them meet farmers who want to work with them, find places to live where they can farm, and raise money to buy land. She laughs, "It's not just weeding broccoli and harvesting beans. Let's go to market, make connections, figure out a product. Let's figure out something you really want to do and expand on that." An example she gives is introducing an apprentice to all of her herbalist friends because he or she wants to be an organic tea grower.

Her view on apprentices is to include people in the farm family and then build on their skills and strengths. She asks, "How can we really facilitate [this person] becoming a part of the organic community?"

Women apprentices, such as Lisa Munniksma, are challenged and empowered. Lisa immediately listed Angela as one of her best mentors. She said, "Every single day on that

Common Treasury Farm's lush pasture helps sustain the farm's livestock, which includes cattle and sheep.

farm, I learned more than anything that I had gotten out of my entire previous farm experience."

Angela gives beginners a chance to take charge of their situations. "I get 'em on a chainsaw, get 'em on a tractor, make them the person who's making the decisions about what we're taking to market and what to sell it for, and give them that power."

She knows this is difficult for many new farmers. "This is not how people arrive here. People find me intimidating for that reason." Her leadership style could be summed up as this: "Hey, you want to do this? Let's go do it. Let's push your boundaries." By the end, hopefully you say, "What boundaries?"

Lisa gained valuable knowledge and practical experience as an apprentice on Angela's farm.

ARE YOU A LEADER?

It's not a far leap from seeking a mentor to becoming one. My first farm-stay experience landed me in a completely different environment from that which I'd come, and after just three days, I was left alone on the farm. Yikes, was I nervous! I tried to do just the minimum work and held my breath and crossed my fingers that there would be no disasters. My prayers were answered, and nobody got hurt and nothing died. Then, more interns arrived. Just those few hours of independence had boosted my confidence enough to be able to show the new interns what the farmer had shown me to do, and soon many hands planting potatoes got the job done. It's a small example of a small step toward leadership but unforgettable nonetheless.

Leadership can take as many forms as there are individuals. The needs that women expressed in our interviews, especially when looking back at their beginnings in farming, always included the critical aspect of hands-on, skill-based leadership. Even if you think you know nothing, you probably know a little more than someone else. As a teacher at a middle school for just one year, I learned how educational it can be to simply give a kid a toolbox and a project and let him or her figure it out. Guiding the students came from laying some ground rules, introducing some concepts, pointing out resources, and maintaining a safe space.

Good leadership is not a line of command, passing instructions down and expecting others to follow them. It is more like a spiral of giving and receiving feedback. Teachers often say that they learn more from their students than they teach them, and the same is true for the mentors we meet in all walks of life. We build skills while the relationship grows naturally. A masculine form of education teaches the feminine how to do something, while a feminine approach educates the masculine on why to do it. This is another way yin and yang can balance and enhance the farming life.

Do you think you have nothing to offer? Think again. Farming in a community of any size requires working with others who have more or less experience or education than you. An attitude of inclusiveness, patience, and teamwork for a common goal will go a long way toward your personal growth and building up your farm's stability.

When I asked Remy about the main challenges she's faced as a female farmer, she replied, "Learning to balance my needs while developing a healthy relationship with others and with cultural norms as I learn what I came here to do in this urban modern reality." Pointing out that she is a leader as a health coach and community gardener, she laughed, "It's so funny, the things we teach are the things we need to learn, right?"

Judy is happy to share her knowledge of animal husbandry with her family's younger generations.

Those of us who do a lot of teaching can begin to feel the well running dry. Replenishing our supply of ideas, insights, and understanding takes time and space. Rekindling our own curiosity about life is a much-needed renewal process. Find ways to be a participant, and you will become a better leader.

I've designed this book to provoke you to dig deeper into your motivations and discover your own resources for accomplishing your goals. Continue asking yourself hard questions and giving yourself time to answer. While you are out in the field picking beans, spreading mulch, thinning radishes, or watering chard, let the ideas germinate and sprout. Give them time to mature, or test them by taking them out on trial runs. Accept feedback. Make modifications. Keep learning. Be your own best student.

What skills, techniques, recipes, methods, or practices do you know well enough to teach to another person?

Describe the leadership or teaching style that feels most natural to you (e.g., working alongside, writing and quizzing, delivering prepared presentations, facilitating conversations, conducting question-and-answer sessions).

In what ways do you currently see yourself as a leader?

In what ways would you like to grow your leadership skills?

Remy recognizes herself as both teacher and student.

YOU AS THE STUDENT

Being a farming student requires a few important things: (1) you know what your needs are, (2) you find out who to ask, and (3) you show up to learn. In the farming world, we are blessed with experts who have tried all kinds of approaches to all kinds of animals and plants. Formal and informal learning each have their benefits and drawbacks. A mix of both seems to be most effective for farmers at any stage of life. Aunt Judy says, "Whatever you want to learn, find someone who's doing it and observe."

Melissa Calhoun studied ecology and environmental science in college. "After I graduated college, I felt I had no skills," she recalls. The areas in which Melissa is self-taught include herbalism, native plant landscapes, giant puppet construction, mandolin playing, singing, community organizing, small business organizing, landscaping and farm development, and interplanting herbs in traditional farming.

Melissa is a great example of someone who recognizes her multifaceted qualities and finds ways to enhance and develop them for a rich and rewarding life. She shares, "I had this memory of being a kid, making weed soup in my driveway. I've always been an herbalist, I just didn't know what an herbalist was." When she met Frank Cook, her first dedicated teacher, he taught her to identify edible plants while on hikes in the woods. She explains, "All he did his whole life was herbalism, giving plant walks and teaching." Learning with Frank helped her find a new perspective on what drew her to plants. "Botany was a language I was familiar with and had learned in college but was completely bored by. Here I am in the woods, and it's all useful."

Unfortunately, mentors are not constant fixtures throughout life. Frank passed away, and, within a year, Melissa encountered a new mentor, Andrew Alexander Ozinskas, who lives in her neck of the woods. He found her through locals who told him to contact her. Now, she enjoys the folk school he leads, and while he does teach workshops, this is not a formal apprenticeship relationship. Rather, Melissa and Andrew go on wildcrafting trips together, and he tells her about every aspect of the craft. She says she can learn much better this way. Knowing how you learn best can be just as important as finding the right resource to teach you.

Jessica Ballard is a good example of a woman farmer who has gained valuable experience with formal and informal training. She did not grow up around farming but began traveling around and helping out on farms as a young adult. After a couple of internships, she decided to earn her degree in sustainable agriculture. While the curriculum was thorough, she says that working on the college's farm for six years really taught her what she needed to know. Networking was also a huge component that she valued, and it led to her current full-time position of comanaging a therapeutic farm for a social services agency. Jessica has some blunt advice that she's learned from her mentors over the years. "I hear people whine about not being able to afford to eat organic," she says. "Go ask somebody with a garden if you can help them. If you want to farm or garden, there are opportunities everywhere. Have your eyes open and be ready."

Melissa learned many farming skills through both self-teaching and working with mentors.

Maintaining relationships is essential because you may not land your dream farm job immediately, and you'll need to build up a network. Jessica says, "Have a goal, but don't be so idealistic that you miss opportunities to show people that you are valuable and useful and you are willing to do what you have to do to get there."

At the same time, be careful how much you say yes to. Check in with yourself and question whether your values are being served. Jessica advises, "It's easy to be convinced that there's nothing out there or to overcommit." She encourages beginners to get used to holding themselves accountable. It's simple, she says. "Be kind to people and do what you say you are going to do." It's easy for young farmers to be such dreamers that they miss opportunities to spend time with elders, teachers, and resources that are readily accessible.

Delia Scott, a very knowledgeable and approachable leader in organic agriculture, reminds us not to overlook good books. "I learned from people who have done it before. I read Eliot Coleman books like crazy, and I think those are probably still the gold standard." Simple but true. Delia repeats the old adage, "We don't have to reinvent the wheel. Others have done this before." A useful tip for searching for reliable information online, Delia says, is to watch for .edu, .org, or .ext in the website's URL. When you enter a search term in a search engine, add ".ext" to narrow the results down to only publications from a cooperative extension service. Finally, she says, "Look around. Ask questions."

Alvina Maynard has proven that by asking questions, you find what you need. She seems very confident, but she shares that it has required effort at times to get past her own assumptions about other people, especially when dealing with experienced male farmers. She confides, "I had in the back of my head that in some situations I'd get laughed off as some silly girl, and I've been surprised. I came out of it with some good, solid mentors."

She has been very pleased with her patient extension agents in Madison County, Kentucky, who helped her navigate grant applications. Alvina says, "I found the processes of USDA funding frustrating." She dealt with a steep learning curve in understanding how the programs worked. Based on her experiences, she says, "NRCS: the 'C' is supposed to be for 'conservation,' but in order to apply for actual funding, you have to have an existing problem, like livestock eroding a hillside. So…it's a restoration program, rather than anticipating issues so we don't have to do restoration."

Alvina with her son; neighbor and helper, Anna; and trusty dog, Ruth.

Lisa Munniksma is an example of someone who figured out what she wanted to learn and then found creative ways to do it. Entering college, she was interested in writing about horses, but her high school advisors told her that she was proficient in writing and editing, and since she was already working for the local newspaper, she should instead go and learn about the industry. This led her to a degree in animal science, but she never wanted to be a farmer. As she traveled and wrote about horses, she says, "I was introduced to many women doing wonderful things in ag on professional and personal levels." Eventually, she took off and shaped her own learning by writing and editing about farming. When that wasn't enough, she became a traveling farmer and writer, filling her life with the dual experiences of formal and informal farm education.

With no prior experience in farming before moving onto a farm in her late forties, Helen Terry looked closer to home for her farm education, much of which has surprised her. "I got to learn about things my husband knew about that I didn't know he knew about."

They keep learning by watching videos on YouTube and talking to local farmers and neighbors. One neighbor in particular has been a mentor to Helen. Catherine Price is a fifth-generation farmer. Helen says, "She's always been someone I could go for a walk with and ask questions, and she'd always have answers."

I began my own garden education by getting dirty with strangers in public. I would show up consistently at a community garden every Saturday morning at 10:00 AM. I learned that when it comes to building raised beds, anything goes, as long as it looks good and holds soil. I learned that building soil means collecting your cardboard boxes and newspapers. I learned that everybody knows something about growing food—and I do mean everybody! Not everybody knows how to hold a spade or cut lettuce, but we all have some bit of garden wisdom.

Delia's own gardens reflect her expertise in organic farming.

I learned that the garden leader at a community garden is more than a leader. This person may set the goals for the day and assign tasks, but the leader is really facilitating a space for learning. The learning could be that my fellow gardener tells me about how to roast beets while I show her how to thin the beet seedlings. It is simple, human, and nourishing to work side-by-side with another person.

The community gardens I volunteered in really helped me out. These gardens were free, U-pick spaces filled with food. There was no pressure to meet any certain demands. Nobody was paying for a CSA share that required a diverse assortment of crops; nobody would be selling the watermelons at a roadside stand; nobody had promised a restaurant a large quantity of a specialty pepper. In short, it was OK to make mistakes. Accidentally sowing *Brassica* seeds way too thickly, yanking out weeds only to realize they were actually carrot tops, mulching heavily over delicate seedlings—I made all these goof-ups and more, and nobody yelled at me. If you need a low-hanging-fruit type of training ground, a community garden is a good choice. Forgiveness comes in abundance, and, more often than not, so does ripe and ready fruit.

Some community gardens run by nonprofit or civic organizations offer formal training programs. Seedleaf, in Lexington, Kentucky, hosts series of classes called Master Community Gardener and Master Community Composter trainings. Similar to the extension service's Master Gardener series, these classes provide guest speakers to lead a variety of classroom and workshop experiences. The topics include soil fertility, season extension, container gardening, mulching, crop rotation, cover crops, sharing abundance, and managing waste.

The benefit of attending community gardening training programs is the focus on very local-specific issues and techniques. They can target not only the broader growing zone but also the microclimates and different soil types of the region. Built into the nature of community gardening or trainings for community gardeners is the opportunity to network. At best, you can find resources that fill a specific need for your garden, like a few more people who would love to show up on sunny Saturdays and dig in the dirt with you.

Lisa enjoys traveling and experiencing different types of farm work.

As a student, it's important to look back from time to time and give yourself a pat on the back for what you have accomplished. It's much easier to see the tangible results of hard labor, but it's not as obvious when you've just climbed a big mountain of learning. When I observed Alvina lead a tour of her alpaca wool studio, she took a question from guest: "What tools do you use to measure the grade [of the wool]?" She answered, "My eyeballs." She then explained that she had taken specific classes to learn how to grade, but she "cheats" with baggies of tested samples that she can compare with her harvested wool. Processing her homegrown fiber was once a subject she knew very little about, but she knew she wanted to learn. Alvina found the resources to learn, got educated, and practiced on her own, and now grading wool is one of her arts, as a result of her training.

What are some major accomplishments from the past year for which you can acknowledge yourself?

What are you proud of yourself for learning in the past year?

How have you educated yourself about a farm-related issue in the past year?

What is one thing you've learned recently that you can share with others? How can you share it (e.g., mentoring, writing, speaking, social media)?

WHAT IS THE COOPERATIVE EXTENSION?

The national Cooperative Extension System celebrated its 100th anniversary in 2014, and with age comes wisdom. The cooperative extension is like a grandparent who is passing down tried-and-true techniques for raising great vegetables. This grandparent also happens to have a Ph.D. in soil science and can tell you what integrated pest management means. The tried-and-true part comes from scientific research, which must be repeatable and measurable. It's less about folk remedies and more about bridging the gap between academic knowledge and practical application.

A law called the Smith-Lever Act, passed in 1914, formalized the activities going on in agricultural organizations since the 1800s. Headquarters for most cooperative extensions sit at each state's land grant colleges. Land grant institutions were established as a response to society's and education's changing focus during the Industrial Revolution and, like the cooperative extension, emphasize practical skills rather than liberal arts. A third partner in the system is the agricultural experiment station, where testing, trials, and field research take place.

The main purpose of the cooperative extension system, as stated in the Smith-Lever Act, was to reach "persons not attending or resident in said colleges in the several communities." In other words, you didn't have to go to college and major in agriculture to glean knowledge and skills.

The term *cooperative* refers to the cost-share structure of extension's funding, of which approximately one-third comes from the USDA and two-thirds come from state and local sources. With so many government entities involved, the system could get bogged down in bureaucracy, but the success of extension lies in the grassroots approach to staffing. More than two million volunteers lead programs by working with county-based professionals.

The six major areas of cooperative extension's focus are 4-H youth development, agriculture, leadership development, natural resources, family and consumer sciences, and community and economic development. Current areas of focus align with modern life and popular trends, including application of science and technology, helping youth break the poverty cycle, disaster response, and the widespread availability of online resources.

The website (www.extension.org) features "Ask an Expert," a space to pose questions that extension agents answer; for example, "Is it safe to put rhubarb leaves in the compost?" and "What is eating my peas?" along with a photo of nibbled pods. Master Gardeners are extensively trained volunteers who answer questions at your local extension office. A quick phone

call can connect you with a real person who will help you solve the mysteries of life, like what animal left its droppings on your porch.

Further, extension.org makes publications easy to understand and easy to find. For example, a quick look on the website under "Small and Backyard Flocks" leads to an introductory article that covers the basics. The "Backyard Flocks" page includes links to other well-respected resources, such as the national sustainable-agriculture program ATTRA, and just another click away is a four-page publication on poultry in urban areas.

You may be numb to the phrase "visit your local extension office" because you've heard it so many times, but it really is the best way to find materials on your specific growing zone and connect with others who can share similar experiences. Workshops and seminars provide a literal and figurative toolbox of skills, knowledge, and materials. I attended a workshop on growing sweet potatoes that included twenty slips (green shoots for planting) to get me started.

I spent the fall of 2013 in the Kentucky FarmStart Course in Madison County, Kentucky. This ten-week series introduced business planning for small farmers of all types. My classmates were beginning farmers who were learning to grow and raise distinctive and diverse products, such as paw paws, alpacas, sorghum, lavender, and even good ol' cattle. We created our own business plans that identified various enterprises and evaluated each one individually to analyze profitability. The experts who taught us came from the local extension office, universities, and the community.

Cooperative extension follows not only an education-based but also a service-based model, adapting to meet their clients' needs. Over half of the world's population now inhabits cities, and we live in an information-based rather than manufacturing-based society. Nevertheless, human relationships and the relevance of educators remain essential to extension's goals.

I live in the city, and my extension office is just a couple of miles away. They have analyzed my soil samples and given me free seed packets. Some of the volunteer Master Gardeners in my area excel at running community gardens. They spread knowledge and share the bounty of their gardens with their neighbors in our city's food deserts. Others are leading the movement to incorporate native plants and pollinator habitat into urban gardens. My horticulture agent spends the majority of her time working with schools to grow gardens, helping pave the way for local farm food to be served in school cafeterias.

Today, the cooperative extension is as active in urban areas as it is in rural ones, but it's still the best-kept secret in many places. Find your new farming ally by using this interactive map to connect with your local extension: http://nifa.usda.gov/partners-and-extension-map.

SECURING THE FUTURE

No matter what you do as a farmer, it's important that you tell others about it. This book illustrates some of the limitless forms that farm education takes. As women, our roles are limitless as well, giving us opportunities to change the shape of the "good ol' boys' club" and reframe the expectations we hold for our successors.

Seeing Alvina Maynard teach college merchandising majors about the concept of slow fashion, it is clear that she cares deeply and is passionate about the quality of products she is creating. She wants that to carry on down the chain of production to the consumer. Passion about a topic can be difficult to define, express clearly, and communicate effectively, especially with skeptics.

On Alvina's alpaca farm, she has found a great point of entry to help newcomers understand the basics. For her tour groups, Alvina starts with a game. She has a set of hand-drawn cards that she gives out for the group to arrange into chronological order. Each card represents a step in the alpaca-to-textile process. The group quickly learned that the dyes for the wool have certain seasons for harvesting (goldenrod in the fall, for example), that you wash the wool after shearing the animals (rather than wrestling a leggy alpaca into a bathtub), and that you must repeatedly sow grass seed to keep the animals fed.

Alvina teaches students about the concept of slow fashion.

Young women in agriculture relate to their educational experiences and use them to help determine where they want to go in the future. They are making an impact just by being present in conversations, enrolling in classes, marketing their products, and speaking up about their abilities. Kaitlyn Elliott is a fifth-generation sorghum farmer who is studying to become an ag teacher. Why not raise sorghum full time? She says, "I'll focus more on gardening when I get older. Sorghum is great, but I probably won't ever get the land, money, and time to live off of it. I'd much rather focus on something that I know will benefit me a lot more." To her, that means teaching and emulating the influence that some of her best teachers had on her.

Sorghum farmer Kaitlyn Elliott.

Recalling how they've been treated by a male-dominated system, the young women I interviewed agreed that they want to see more equal opportunities in the public schools for classes such as construction and welding. Kaitlyn took some of these classes and was definitely a minority in each one. She remembers, "It was a good class, but we weren't graded equally. We were automatically given an A because we weren't going to be as good."

Marlen Hammond comments on the culture that surrounds high-school shop classes: "Classes should be required instead of optional, so you don't feel like you're the only woman in the class. Everyone should have some kind of mechanical training [so they] know how to change a tire at least."

Marlen's colleague Lizzy Guthrie proves that women can figure out mechanical problems. She says, "I remember being broken down in the desert once, completely stuck, and I could not figure out how to get things rolling again. Eventually, I just laid underneath the tractor and read the owner's manual, and I figured it out. Being put in those scenarios, you don't really have another option, but it would have been nice to have had specific training on how an engine works."

Kaitlyn is young, but she's been working with her grandfather at farmers' markets long enough to know what the public perception of a female farmer is. She says, "If you have a business with a man, they never refer to you as being part of it. They ask how your dad's business or your granddad's business is, not taking into account that you actually do work. If there's a man, there's no respect given [to the female]. With sorghum, a lot of people just assume my role is to bottle it and label it. I guess I'm supposed to be like a little housewife. In reality, he's kicked back in a chair while I'm out there with a machete, cutting the stuff down." She said that the people with whom she works closely don't discourage her like those who see her only at the market. "It's the people who don't see you working who make it harder because they don't think women can do that type of work."

Women have to be prepared on some level to intercept, deflect, and overcome stereotypes. Gender studies specialist and Ph.D. student Katie Ratajczak empathizes with the effects of being pigeonholed with assumptions, and it affects both genders. She asserts, "If you're interested in farming, not because you want to be the breadwinner and you're working with the ground as a 'manly' thing, but if you want to think of it as becoming in touch with the land, working with it, and returning things to nature, that's [got] a really feminine orientation to it."

Lisa Munniksma has had mixed reactions from men she interacts with when going to markets in cities like Cincinnati and Louisville. She's heard the surprised reaction of "Oh, you actually grew this?" and other humorous comments. She tells about a time when she had horses in a half-ton truck and gooseneck trailer. She says, "I pull into the Flying J truck stop, and this guy says, 'That's a big rig for a little lady.' It just made me want to punch the guy in the face."

It does depend on the person's intentions, though. Sometimes, a stranger's comments can actually be encouraging. Lisa recalls a time when she was leaving a market, driving a fifteen-passenger van with a trailer attached. "There was an older man who said, 'Tell me you're going out there to drive that van and trailer.' I told him, 'Yes, sir, I am.' And he said, 'You go get it!'" Lisa smiles at remembering the small boost of encouragement that came her way rather than a condescending attitude.

Lisa laughs about having male friends who call and say, "Hey, I have to do this thing with this trailer. Can you come over and drive this for me?" So, how did she learn to drive big things? She credits two horse trainers, both female, where she boarded her horse for patiently teaching her how to drive and back up trailers.

Joscelyn Strange had male mentors in her family, including her dad; in Future Farmers of America (FFA); and in 4-H who taught her about dairy cows, and her mom showed hogs. She says she learned more from men because they were doing what she wanted to do, and she never heard negative comments from her family members about being a female. As she developed her niche of showing cattle, she has seen a big influence from women in the industry. She says, "I see it all the time because I go to dairy shows; nine times out of ten, the people on the halters are women." Joscelyn is pursuing a degree in animal science. True to her upbringing, she plans on running her own cattle operation and preserving the Ayrshire breed with which her family works intensively.

Great leaders bridge a gap. They illuminate a path that gets you from point A to point B. I often hear the phrase "meet people where they are" from leaders I respect. This openness and adaptability can work wonders with those who are used to more dogmatic leadership styles. Do your best to find common ground with colleagues of both sexes and all types.

Changing the culture, changing attitudes, opening up minds, and shifting expectations can really only happen on personal levels and through connections. The women I interviewed all expressed a willingness to work with the masculine and patriarchal systems, as well as individuals, as long as those men were willing to work with the women and hear the women's point of view as well. I'll end this chapter with a quote Alvina Maynard shared with me, something she heard at an Empowering Women Veterans conference: "We cannot influence change by damning our neighbors."

Chapter 6
Lessons Learned

Warning: This chapter is about the worst of times as a woman farmer. These could be sudden, devastating disasters or a series of small, undermining issues that add up. We'll boldly look these monsters in the eye and find out what stories they are trying to tell us.

What happens after harvesting and sharing in the natural cycle of life? It's time to let some things rot. In a perfect system, nothing is wasted. Nitrogen, carbon, and wisdom from experience continue to feed the farm long after the pumpkins have been picked. It's time to put the garden to bed, allow cover crops to take over, or let the fields go fallow for a season. When it's time to rest our bodies, our minds can contemplate the memories and notes from the past months or years.

Compost isn't always pretty, yet there is beauty in its alchemy. Watch worms and other tiny life forms turn the used-up remains of a season's work into rich, fertile soil, preparing the way for the seeds we've been saving.

Farm care revolves around life and death, and the horror stories are bad enough to turn many away from farming. A very human reaction is to try to forget the trauma and move on, but to really progress, grow, and prevent the same occurrences in the future, we are better off in the long run if we choose to take a good, long look at the worst of it and let the hard times teach us.

The women farmers over the age of fifty with whom I spoke echoed each other's thoughts about what resources they wish they'd had when they were starting out. For example, one woman remarked, "I learned a lot about a lot of things through experience, but it would have been nice if someone had told me some basic facts early on." The women in their forties generally had to work pretty hard and also figure things out for themselves, and they openly acknowledged the need for a community.

The women in their thirties made many decisions to engage more deeply with the natural world and nourish their own bodies, in turn gaining much knowledge about self-care and becoming great role models for young women. The twenty-something women farmers don't necessarily know it all, but they are confident that they can find what they need, with the exception of land.

What are the top three worst experiences you've had as a farmer?

What did you learn from these experiences?

What are the top three worst-case scenarios you can imagine for your farming future?

How can you prepare for each of these scenarios? List your steps and check them off as you complete each one.

Aunt Judy's words of advice drive home the point: "When you raise a garden or animals, because they're alive, there are variables. It's not *if*, it's *when*. How am I going to handle this when that comes up?" She says that farmers are smart because we are always looking at solving problems, and that's part of life.

REFLECTIONS

I asked Remy Hendrych why she chose the topics of composting and poop as conversation starters in urban ecology. How does this relate to her community leadership and education efforts? She smiled and replied, "Well, luckily compost only takes three to six months, right?" Seeing the effects of natural cycles usually takes much more time than we have the patience for. But watching something decompose stirs up all kinds of questions about our span of life on this earth and what will happen after we are gone. Not to mention the value of the fertilizer itself. Our systems thrive on by-products.

Helen Terry reflects on two points: what she wished she had done differently in one situation, and a risk she was glad she took in another situation. Both of these situations involved getting help from others, something we all struggle with from time to time.

Helen and her husband, Joe, live in southern Texas, where the warm climate can support year-round fruit production. They enlisted the help of someone to plant their orchard, but that person hadn't planted the trees in an optimal way. Helen says, "At the

Remy connects with the spiritual side of nature's cycles.

time, I didn't know enough about fruit trees to know what their needs were. Now, years later, I can see that they could have been planted better. I always feel like I'm one step behind in my learning curve."

Another tree situation involved a massive oak tree. They thought that it was dying, and a tree expert from Texas A&M University came out and recommended a $500 treatment to lengthen the life of the old tree. Helen says, "It was a big risk. I didn't know if that man was conning us, and if it was worth spending $500 on a tree when we were on a very tight budget, again being in the dark. That actually did end up working out great. We're really glad that we did it, and the tree is blossoming."

SHIFTING PERSPECTIVES

Katie Ratajczak has studied violence against women. She says it is a pandemic in the United States, particularly on college campuses and among migrant farm workers. Women in migrant farm-labor situations feel trapped because if they expose the problems, they could get deported, lose their jobs, and be separated from their families. It's a precarious situation that puts them at a high risk.

In Katie's feminism studies, she has learned of the twisted link between women connecting with the earth and being forced to stay there. "Society has tried to position women closer to nature so that they can't reach the levels of rationality and objectivity that men can. By keeping women tied to nature and to things that seem to be out of their control, they can't enter into this 'higher realm' of scientific rationality," she says. "Women may be in tune with cycles, but that shouldn't be used as a way to keep them in a subjugated position."

What has been perceived as "progress" as nations become industrialized—Big Ag attempting to control seeds, and wars waged in the name of capitalism—is directly impacting women, particularly women who are closely connected to the land. Katie points out that the social movements in our nation's memory, such as women's right to vote in the 1920s and war protests in the 1960s, began with those who we might call the victims standing up and fighting back.

Katie explains, "People who are being affected say, 'Hey! Wait a minute!' What's happening isn't benefiting them." If you are not suffering, it's not in your self-interest to point out the problem. Katie concludes, "Get women involved in this process, and it will help point out some of the problems and push us in a better direction."

Jessica Ballard and those who work in gardens that focus on providing healing spaces are continually reflecting and adjusting to meet the needs of both the clients and the crops. At GreenHouse17, where Jessica is the farm comanager, the shelter's mission is "nurturing lives harmed by intimate partner abuse." The former name of the organization included the term *domestic violence*, but Jessica and the farming program facilitated a shift in thinking, which is reflected in the new terminology. *Victim* is replaced by *survivor*.

The symbolism is more important than the farm's actual products. Jessica describes it: "A greenhouse is a protective space for organisms, flowers, for when the world outside is not as hospitable, when you need a space to get a jumpstart. That's the type of space we want to hold as an agency." Within the safe space, the nurturing circles take many different forms. Jessica says, "Now we have craft groups, a sewing group. We make products, have book clubs, do journaling. It's a different way of holding space with survivors of trauma, so it's not like 'Let's talk about how you were violated.'" Jessica and her colleagues allow healing to happen naturally, in natural environments. She explains, "The farm started as 'Let's hang out as women, and this is a safe space, so let's do something we love or that is challenging.'"

Jessica further defines how the greenhouse and farm support healing. "Clients I'm working with, in particular, are coming from a place of hearing men say, 'You can't, you won't, you wouldn't know how to do that.' But really, you can do this!"

She emphasizes that the agency supports a whole mental paradigm shift that has illustrated the importance of having a place for women to work together. She says, "There's something intimate about working with other women. I am a spiritual woo-woo mama, and I know that nature is working some magic on us, even when I'm grumpy. Because we are birthers, we are creators, it is a little more acceptable for us to be vulnerable and emotional. As the earth is doing its work on us, just from a healing perspective, we're just a little more open to allowing that, even people who are hardened and never want to break open. There's a little more potential for that healing force to come through when it's just women working together. It's really special."

Jessica reflects on what this farming experience has shown her about her own path. "I always see myself working with cycles of nature. Knowing when to plant each seed—that's my code. I couldn't really imagine at this point not doing some kind of farming or gardening." She foresees a time when she will have to shift gears due to the physically, emotionally, and spiritually demanding nature of her work.

As Jessica watches the dynamics at play among the cycles, the women, and the larger world, she says, "Sometimes, the plants are just in the background." She has observed the male energy of housing construction disturbing the sanctity of the garden and exercises her communication and compassion skills while clearly setting boundaries. In these times of transition and upheaval, Jessica notices her protective mama-bear instincts rising. She recalls, "The plants have had more issues; it was a bad year for Japanese beetles. I can't tell if there are more issues or if I'm just looking closer than before." Being sensitive to the uncomfortable situations that arise can be a mixed blessing. Jessica says, "We are all sensitive. If we don't know we're sensitive, that's when it bites us."

Melissa practices chemical-free farming and is attuned to the cycles of the natural systems around her.

FINDING COMMUNITY

Melissa Calhoun reflects on what resources she still needs today, and she realizes that her needs revolve around finding like-minded individuals who share an interest in thinking outside a prescriptive box. By observing the natural systems around her and working with nature rather than trying

Rachael gently taps her beehives to listen for activity inside.

to conquer it, she finds herself on the fringe of even the "green" communities. She explains the community she is interested in building. "For instance, folks who can speak knowledgeably about and find alternatives to dependence on pesticides, herbicides, and invasive species eradication, which we consider genocidal and xenophobic warfare. We observe this warfare largely practiced in service to a paradigm that harms rather than serves us and the earth. Most with power or ownership over land are insisting on the belief that these practices are helpful to the earth and ignoring the lack of actual research on the effects of such practices…that excludes observation of the natural succession of unmanaged forests and other landscapes."

She doesn't seek to distance herself from existing landscaping and conservation communities, and she acknowledges the complex economic and job-security motives behind "spraying routines, which no one bothers to mention are infinite in scope and endless in goal as plants and insects continuously adapt and attempt to fill their nature-given roles." Melissa has come to her own conclusions through her direct experience with natural cycles. She concludes, "That's probably my most passionate soapbox as I encourage others to rewild and rethink the stereotypes we've been given."

Lisa Munniksma recalled some of her worst times as a female journalist covering agriculture topics. She said a low point was at a national farm-machinery show. She remembers, "It was soul-shredding. Why am I even here as a woman? Nobody is even taking me seriously." She avoided talking to anyone she didn't have to, and she dealt only with certain representatives. She says that their general attitude was very condescending and that they "mansplained" everything. I asked Lisa what her definition of "mansplaining" is, and she replied, "When men explain the most mundane details of whatever the topic is to women because us poor women just don't know any better."

Rachael Dupree was surprised by the ups and downs of the first few months on her farm as she went through the transition from living in the city to the country, of getting

used to the new home, and of feeling challenged to find personal connections with the community. But it is still all new, and with time, she believes her impressions will change. Nevertheless, she is learning about herself in ways she didn't expect. Rachael shared, "Biases are coming out that I didn't even realize I had, so that's humbling. I kind of think highly of myself, and I shouldn't. A lot of different thoughts and emotions are happening. I figured I'd move here and have to work hard, but there are things I just never expected—the adjustment. I wanted to do this, but it's culture shock."

Rachael blogs about the process, some of which is just dealing with bugs and ticks and mowers breaking down while planning a wedding and working forty hours a week. She says, "I go through stages. I'm super-excited, and then I hit a wall, like, 'Oh my word, should we have done this?'" The uplifting aspect is the thought of opening her farm to others. She looks forward to hosting, sharing, and teaching with all that her fifty acres has to offer.

BOUNDARIES

Aunt Judy reflects on some of her earliest experiences as a young girl and what they taught her about handling animals. Judy was a brave, untamed child herself, and she loved the summers on her uncle's farm. One summer in particular, she was looking forward to seeing a yearling colt, and she excitedly went out, jumped on a horse, and started working to separate the mother from the herd. "This yearling colt," as Judy tells it, "came straight toward me, so I thought she was sweet and friendly. She came straight at me, turned around, and lickety-split! Kicked me right in the ankle and broke my ankle."

Two of Judy's happy herd.

Judy tells this story as if it happened yesterday, and she remembers the shock. "I didn't see that coming at all, and I learned something there: the colt didn't want the mom to move. I was interfering with their whole thing, and I didn't know that. I found out the hard way."

Judy teaches others what she learned that day. "All of nature works on boundary systems." Chickens, horses, dogs, preschoolers—any living thing is going to interact somehow with another. Working on a ranch, Judy saw the dynamics play out as boundaries were established, limits were tested, and lines were drawn. She learned with horses how to handle it. When a new one arrives, she says, "You put [the new one] in a paddock next to the herd so they can do all this fussing across the fence. It's not as dangerous. It gives them a chance to get familiar, and they can blend better when they get together."

REALIZATIONS

When I asked female farmers, "What are some strengths or natural qualities that women farmers bring?" Angela Wartes-Kahl's answer summed it up. She responded, "Ability to persevere in the face of opposition and adapt to new conditions."

Alvina Maynard hasn't shielded her seventeen-year-old farmsitter from the realities of raising livestock. Anna, who lives on the neighboring farm, takes care of Alvina's farm when she goes out of town. And, every time, baby alpacas are born on Anna's watch. They laugh about it, and Alvina jokes, "They'll be overdue, and we're waiting for a baby to come, and we have a trip planned, then Anna takes over—the first year, there were three calves born on her watch!"

Angela transports compost on the farm.

The conversation takes a turn as Anna shares that, sadly, there was one that didn't survive. Alvina is quick to point out that it wasn't Anna's fault. "Anna beats herself up a lot about things like this." Anna replies, "Well, I felt bad. It's a baby, you know."

Domesticated livestock face a different sort of danger than their wild cousins. While they are housed and fed and looked after, they have been bred to depend on humans so much that they've lost some of the instincts that keep their young protected. Another realization is that we humans are responsible for putting these

animals in a vulnerable situation, and we have to figure out how to be their protectors, their herd.

The animals on Alvina's farm are much more than food and fiber factories. They provide additional services, and only after years of living in harmony are the farmers able to take a step back and recognize the animals' real value. While we laugh about Alvina's goat, a.k.a. the "weed-eater," sitting on top of rolled-up fencing, Alvina's big white farm dog, Ruth, runs out into the road and stops a FedEx truck from making its deliveries. This dog not only shows the driver who owns the road, but she actually herds the truck onto a side road.

The entertainment value is priceless, and then Anna tells the group of visitors a story about Ruth. The dog had woken Alvina in the middle of the night, barking loudly. There was no predator, no intruder. Ruth led Alvina to the barn to find that a water main had broken and was flooding the barn. Her canine alarm is invaluable in many ways and an important part of her farm's community..

CONFIDENCE IN COMMUNITY

Finding community is a recurring theme, one that Lisa Munniksma mentions as she discusses what she has observed about men and women as activists in food movements. "Looking at the sustainable food and farming side, I found that men were more likely

Lisa sells a variety of fresh produce at a farmers' market.

to be involved in the activism part than the community-building part. [Activism] is also really important work, and there are tons of women doing great activism work, too. That's important, because it makes a difference, but so does the community-building part."

Lisa compares the equestrian community with agrarian communities. "In the horse world, you meet up with friends, go show, and hang out at the barn. It felt different than the farm world." She observes that when you're growing food for yourself and others, there is a need to come together and get work done. She says, "Obviously, this is difficult work, and there's so much to be learned from other people. Especially in the peak of summer, when it's so stinking hot, and everything needs to be done yesterday. You have stuff to plant, stuff to weed, stuff to harvest, and stuff to preserve—right now. It becomes apparent really quickly that you can't do this by yourself."

Lisa refers to the Greenhorns book. She paraphrases, "You're going to be in the field for long days, in the sun, in the rain, when you're not feeling well, when you're sad, when you're happy, and you'll be experiencing all these emotions. Sometimes you'll be alone, sometimes you'll be with people you love, and sometimes you'll be with people you can't stand." Lisa continues, "All of this is really going to shape you into a different person or bring out your true character."

She has worked in many different configurations of farming with interns, as an intern, and supervising interns. "They aren't really prepared for the personal transformation that takes place," Lisa notices. At the farm where she landed a regular job, Adam Barr's farm in Kentucky, she realized, "I felt a little like I had cheated that system, because by the time I got to Adam's, I had been traveling solo for two years and, boy, I had worked through some stuff."

She knows it is different for every individual, and some just push their way through it, but Lisa says that is cheating, too. Lisa encourages herself and others to give themselves grace to feel whatever comes up. "This is part of it. Working with the land is healing."

We do not understand many of the difficulties of being a woman farmer while we are experiencing them, but with some distance and space to figure things out, we can see the situations more clearly. Many of us feel a vague sense of discomfort working with men but have trouble identifying it. When we get together, we don't bash men, but we do begin to recognize some of the common denominators in our stories.

Marlen Hammond uses the phrase "unconscious discouragement" to describe the effect that some men have on their female coworkers. She says, "When women try to do their jobs, even if it's not on purpose, [the men] try to take over. When you're doing something, and you think you're doing it well, it's very discouraging, and that makes it harder."

REDEMPTION

The most dramatic example of learning through mistakes comes from Aunt Judy. She begins the story by saying, "The first thing in nature's cycles is to have a devastation." The family of eight had just moved to West Virginia, where Judy's sister also lived. Judy had been raising goats for her family's milk but hadn't really made it a viable business. She accepted a proposal from her sister to help her get more goats for breeding. A combination of tragic mistakes led to an "abortion storm," the term used by a dismayed veterinary specialist to describe a massive number of miscarriages.

What went so horribly wrong? As Judy tells it, "The herd was supposed to be all unbred does [but] turned out to be all bred nannies. They came in toward the end of the year, coinciding with a horrible snowstorm. The fence wasn't finished, so we had them in smaller lots, and they were all fighting." Plus, the goats' origins were unknown, which normally would mean putting them in quarantine until they proved to be healthy and disease-free. These goats were neither healthy nor disease-free, and they were all intermingling and fighting while pregnant. Add to the mix naive new farmers and a tenuous business relationship based on a significant investment by a family member, and it was a recipe for disaster. They lost about 75 percent of their herd by the end of it.

Amazingly, Judy found redemption from the biggest mistakes. "That's where a lot of life lessons come from. Things we go through that are just horrible. I remember it like it was yesterday because it was so intense. The kids remember it, too. We had never seen so much death. It could have damaged our dream. It looked like the dream was devastated, but I saw such a positive principle in the aftermath."

Judy has persevered through the challenges of farming to share her knowledge with her children and grandchildren.

As they were grappling with how to proceed, Judy says, "It was hard for me to see hope, to see what the positive was, until [someone suggested] that we needed to eradicate them all." She instinctively knew that was not the answer.

Judy went out and observed the remainder of the herd, about thirty-five surviving goats. "I was watching the goats while I was thinking through all this stuff. I watched them playing, and moving, and the way that they were. They were so solid. There wasn't a one of them limping, acting like they were sick. And I'd just gone through all the sick, so I was so observant about anything that might possibly look like a

Native Kentucky grasses.

problem. There weren't any of them doing it." She calls this her "gold." "The remainder, the remnant, had been through the test. We had the gold that would make a foundation herd that would [be] better in the long run than anything we could have done."

This lesson has remained with her, and she ponders it deeply. "I saw something then that I hadn't seen before in that dramatic of a way. The surviving thing. Plants are the same way. Something takes them all down, but the ones that survive are the ones you want to keep the seeds from because they're the ones that are the most resilient. Science has proven that now."

She also realizes that this manipulation of nature is avoidable. "If we hadn't done all those things that were unnatural to those goats that were brought in, if we hadn't put in all those other factors, we wouldn't have had that abortion storm. We had all those things that weren't natural together in one place, and it stormed." She refers to this as a process that weeded out what had to be removed. "The worst time was the best time. It opened my eyes to how things work naturally. No matter how much control we think we have, it may not be what we need. [The experience] gave me a glimpse of how sometimes the remnant is your gold. It's the place where you need to focus. The pedigree is not the gold at all."

MAKING DECISIONS

Choose your battles, decide where to put your energy, and make the most of your resources. Alvina Maynard is a tough lady, but she doesn't have to prove it to everyone.

I asked if she'd had any unpleasant experiences that she related to being a woman farmer, and she replied that she's not sure if it was due to being a woman or just being an outsider.

Rural Kentuckians are very rooted in their communities, where many generations remain. Alvina moved to Kentucky from California as a young woman, and she felt like she was in a foreign land. As a military veteran, she knows what that feels like. She said, "I have had some occasions where I send my husband in…and decided not to be a customer of certain feed stores. I was just not treated equally as a male counterpart. I can go to a different feed store. There's no reason to give the business my money in an area where there are other options. Thankfully, I've not been in a situation where I've been out of options." Sometimes, she just enjoys standing her ground and showing the good ol' boys that they don't have her all figured out. She adds with a mischievous smile, "I do think it's fun to dress up and wear high heels and pick up a fifty-pound bag of feed and sling it over my shoulder."

After being involved in several different farming scenarios, Melissa Calhoun has come to some conclusions. She appreciates that lots of her friends are great farmers, and she has plenty of access to food. She is happily releasing that aspect of being a farmer. She finds fulfillment in growing herbs exclusively. She declares, "I love deciding that I'm an herbalist!"

Aunt Judy's reflections harken back to forty years of farming with her husband and raising six children and the decision that prompted them down that road to begin with. "We thought we'd get real freedom. I think we can get a piece of the freedom. But I think it takes generations." She contemplates this idea of bondage that holds on for generations. She says, "I think you break it, but it's only the beginning of complete freedom. It may not show up in your life." Judy refers to the biblical example of Abraham, who had given up on living to see the prophecy play out that he would become the father of many nations.

LIVING FULLY

"The tough stuff has made the sweet things all that much sweeter," writes Rachael Dupree on a social media post. This newbie farmer practices reflection and appreciation of the rough times. Living fully means embracing all that comes our way and taking time to explore the deeper meaning in the compost pile.

Remy Hendrych muses on the idea of decomposition as being deeply magical, and that compost itself is a small-scale version of what we're doing with our lives on a larger

scale. "We're decomposing old structures and old systems. We're the fungi, bacteria, and invertebrates of society, eating away at ways of doing things that aren't really serving us. We're decomposing. We won't get to see all the fruits of that during our lifetime. We won't get to see that juicy black gold. The societal version of that would be a nature-connected community where there are people of every age group living together, celebrating, having rites of passage, growing their own food, helping others—that would be the black gold of our time. We get to see that on such a small scale, taking waste that we'd just put in the landfill and throw away like it's nothing, like it's not sacred. I get to go around the city and collect it twice a week. It is very meaningful to me to take that symbol—why do we even call it 'waste?'—of our times that we are a consuming-focused rather than creating-focused culture."

Aunt Judy believes that decomposition reveals what is important. "Pain is one of the best teachers we have. Why does pain have to bring the necessary sight? Because it gets your attention? The more I run from it and hide from it, the less I'm going to know." Judy has come to this conclusion: "Part of our life on this earth is just growing up and getting in the groove with what's already set in place. We think we know it all, and we can just plow through. We are not the God of the universe. That's what it boils down to."

She continues, "What do I know and what do I not know? The older I get, the more I find that I don't know. The more I connect with the principles that are already set, the better things work."

Take time to reflect on the farming experiences you've had over the course of a growing season or a calendar year. Summarize the highs and lows and what you've learned. If you are ready, share what you've learned with your family, friends, farm supporters, or customers. Look back at the lessons and glean the gold.

Chapter 7

Integrating Farm and Life

Feminism, as the mainstream, middle-class American knows it, is about women wanting equity in the workforce. This enables women to have their own money, own their own property, keep their own last names, and be their own people without dependency on a husband. Katie Ratajczak points out that this is not the dream of women worldwide. For example, Katie says, "Women of color have been oppressed and been in the workplace all their lives. The portrayal of women in the home as stuck, fragile, and oppressed is not their experience." Taking into account that the assumptions about women's roles described in this book reflect the viewpoints of the women farmers I've interviewed and my own experiences, I intend not to exclude anyone. However, I am limited in my own perspective and cannot account for all sides of the topic.

I hope that this book provokes conversations about what it is that you, a multifaceted human animal with instincts, intuition, intelligence, and spirit, all of which hold masculine and feminine aspects, wish to offer the world and what you really need to support those gifts. Retreating to your dream farm and living off the land could be the path for you, yet most of us will integrate with mainstream society to make the best of both worlds. In this chapter, we'll explore some ways in which integration supports the dream of many women farmers.

YOUR CAREER

What feeds you? The lifestyle that feeds your soul and the job that puts food on the table might seem like opposite poles. Farm life and other duties, like a nine-to-five job, can be integrated as a whole system rather than separated into distinct parts.

Many farmers get very creative with their career choices—like a self-designed major in a college where the exact program for you is not offered. We are lifelong learners. Design your own unique version of the opportunities and shape them to fit you personally.

Lisa Munniksma defines her career as freelance farming and writing. I ask if the term *freelance* applies to both aspects, and the answer is yes. Her farming and writing careers are intertwined and both are opportunistic in nature, leaving much to chance. Flexibility and freedom suit Lisa. "I write about food and farming. Everything I do on the farm impacts my writing, and everything I write about, I take back to the farm."

Angela Wartes-Kahl and her husband, Garth, run a farm and are also organic inspectors and consultants, telecommuting from their rural home. Melissa Calhoun works her day job at a family-owned tool supply business, farms at home for fun, and finds ways to nurture her creative pursuits as well. This is a new balancing act for her as she transitions from sharing a mortgage with four others to now handling it alone.

Alvina Maynard presents another creative approach with the support of her husband and his off-farm job. She keeps her two-year-old son by her side while she manages the alpaca-farm business, and her six-year-old daughter helps out when she's not in school.

The other support is her military job. She is in the Air Force Reserves after having served six years on active duty. Her job as a reservist has been in federal law enforcement. Alvina is gone one weekend a month and two weeks a year, but she manages to keep it all in balance. She says that the military training has benefited her farm business. She says, "My primary skill set in the Air Force is to talk to people and build relationships. This is invaluable as a small-business owner—networking, building relationships across the community, and building my small-business customer base." In other ways, the military prepared her for the unpredictability of farming. It taught her the importance of flexibility and the ability to adjust to the circumstances. "In both of the jobs, God is just up there, laughing," Alvina says. "Everything goes haywire, and you either laugh or cry, adjust, and figure out how to overcome the challenges."

Melissa works for a tool-supply business and uses her tool expertise on her own farm.

RURAL AND URBAN FARMS

Whether you count your acreage by the hundreds or by fractions, there are pros and cons to growing food and raising animals anywhere. The picture-perfect ranch house or cabin nestled in the woods might be great for giving kids and livestock the space they need to frolic, but for some people, the isolation and disconnection from a bustling neighborhood can be too depressing. Growing food in the inner city is a great way to meet many social needs for yourself and for folks who need fresh food, but land is at a premium, and you may always be in competition with development.

Lizzy Guthrie's parents were the first generation to not live in the city. Her grandparents hailed from Boston and Cincinnati. When Lizzy's mom and dad got married, they bought a farm. Living in contrast to these generations' typical patterns, her grandparents have a small vegetable garden in the suburbs, her parents hold onto acreage, and Lizzy hopes to move out West to gain even more space.

Delia Scott grew up on what she calls the "classic Kentucky farm" on five acres next to her grandfather's hundred acres of tobacco, soybeans, and cattle. She didn't appreciate the country life then. "I just wished I had people to play with," says Delia. "I had to move to Lexington to go to school to find a love of agriculture and horticulture. The joke in my family is that I had to move to the big city to learn that I wanted to be a farmer." She worked as an extension agent, providing support for many urban farms in Lexington before becoming the executive director of the Organic Association of Kentucky. Having seen both sides, rural and urban, Delia hopes to return to the country, although she is an urban farmer for now.

Agriculture is a full-time job for Marlen.

Marlen Hammond is considering what her farming future will look like. Marlen currently lives in town and doesn't feel the need to garden after spending forty hours a week doing it for a paycheck. She is a young mother working at an agritourism farm, where she has an enjoyable, stable, and diverse position with benefits. Her mother, Margrit Ciccarelli, will someday give up the family's 500-acre dairy farm, and Margrit hopes it will go to one of her daughters. Marlen is the obvious choice because she is the only one involved in agriculture. Time will tell if she carries on the heritage.

Angela Wartes-Kahl's farm is situated in a beautiful location with a south-facing slope, near a creek, and surrounded by national forest. However, she says, "We need a source of affordable labor, but we are too remote." The rainy winters of the Pacific Northwest regularly cause an exodus of interns and summer help, and Angela must travel more than an hour to get to any of the area's profitable markets.

Urban and suburban farming hold much potential, including great educational opportunities, by meeting people where they work, live, and go to school. Suburbia brings an infrastructure of irrigated green spaces that can transcend the standard turf grass and could become swaths of pollinator corridors, colorful wildflower gardens, or interconnected community gardens. Urban farmers learn to work with existing structures and integrate small solutions at the local level.

Remy Hendrych started her urban farm, Urban Indigenous, along with a small group of people who created an intentional community in downtown Lexington, Kentucky. A Hawaii native who left a corporate career to WWOOF on an organic farm in Mexico, Remy is dedicated to the mission of nourishing her community. "Urban farming was why I came to Lexington," she says. "I wanted to live in community with people who wanted to raise their own food and experiment with permaculture. That was the impetus. Now I'm seeing it was just a tiny piece of a greater puzzle of what it means to be a human being [who is] alive and choosing to live in an urban environment."

The residents of her household community are not deeply involved in the gardening aspects, yet Remy sees the effect that the food forest has on her housemates and neighborhood—physically, emotionally, and spiritually. She says, "The practice of urban farming can meet a lot of needs if you recognize it and show up each day. It can be just meeting your food needs, or it can be a lot more."

What this means to her house at this time is inspiration. She hosts gatherings, often impromptu, sometimes planned, such as building an earthen mass heater, creating a *hugelkultur* (raised bed) installation, or gathering a salad. Remy describes what happens: "A bunch of people will join, and it will be a moment for others to say, 'Wow, I want

more of this in my life.' It's not yet a core practice that is being used in conjunction with the other goals or the spiritual path that many of us are walking. It's not being actively done by anyone else but certainly a piece of inspiration and aspiration for all the residents involved, either by getting their hands dirty or watching a garden grow day in and day out." Remy isn't discouraged by the noncommittal attitudes of the residents. She calls it an "incredibly fruitful path."

What Remy is doing in the city is a very natural type of gardening, with layers of food-producing plants that are relatively low-maintenance perennials. However, this doesn't conform to the standards to which most urban areas are accustomed. For example, in some cities, the zoning laws pertaining to plant height and composting practices conflict with urban farming. However, as we all become more aware, more educated, and more compassionate about the interconnectedness of all life on earth, we will see these walls begin to come down.

While America's major cities, small towns, and suburbs are beginning to support the local-food movement, there are also major strides in cities worldwide. Take a look at some of the urban-farming initiatives starting from grassroots ideas as well as governmental support. Maybe you can find inspiration to jump-start your own urban farm or organize interest to bring fresh food into your community. Some countries need urban gardens simply for survival; others need them to push technological boundaries. Take a tour of five of these international urban-farming initiatives.

ARGENTINA

Brazil's southern neighbor has made an impressive recovery after suffering an economic and social crisis in 2001. In 2002, the municipal government in the Argentinian city of Rosario, in the province of Santa Fe, launched an urban agriculture initiative, relying on community partners to supply twenty community gardens with tools and seeds.

A fruit market in Buenos Aires.

Through teamwork, the community well outgrew the original plan, implementing 800 community gardens that feed 40,000 people. Tiny farms continue to pop up throughout the city, even on former garbage-dump sites. Farmers plant cover crops, mulch with wood chips, and add compost and manure to remediate heavy-metal contamination and build fertility in the soils.

As part of the government program, five *parque huertas* (garden parks) are connected to markets throughout the city. These unique green spaces combine cultural, sports, educational, and horticultural activities, leaving space for commercial growers as well as home vegetable gardeners.

The key that makes Rosario's urban-farming initiative a winner is the good game plan: integration of infrastructure with community needs and cultural strengths. Not many South American cities have policies that support urban agriculture to this extent. The original motivation for a public-works project was to relieve an economic crisis, and the emergency needs were met while establishing agriculture permanently in the city.

SOUTH KOREA

Seoul is a city of ten million people buzzing around like a beehive full of tightly packed, orderly workers. But bees need flowers, and even in this metropolis, plants are thriving on balconies, on rooftops, and in tiny spaces. Seoul's city hall has a beehive on its roof, and young entrepreneurs run an urban apiary called Urban Bees Seoul.

Gardening, especially in community, is nothing new to Koreans. About one in seventy people in Seoul gardens. The agriculture ministry is pushing for participation in urban gardening to reach five million people nationwide by the year 2020. The ministry also plans to repurpose 7,200 spaces, including undeveloped land and unused buildings, into gardens.

Green space in Seoul among low-energy residential buildings.

The cultural norm in Seoul is to use resources efficiently and to value high-quality produce. Even the mainstream big-box store Lotte Mart has a new urban-farming gimmick: growing hydroponic lettuce and other vegetables right inside their stores so customers can be part of the process and know where their food comes from.

The biggest innovation in South Korea, however, is vertical gardening—matching the city's architectural appeal and limited ground space. Mimicking the ancient practices of terracing, architects from around the world have designed closed systems of high-rise diversified agriculture, managing conditions for greens, fish, chicken, and even cattle to grow indoors.

THE NETHERLANDS

The Netherlands is also thinking vertically about urban-farm design. Despite limited land, technology is abundant here, and they toss out the theory that all you need to grow food is sunlight, water, and soil.

It's not uncommon to grow beans, cucumbers, and strawberries indoors under LED lights, a system that yields three times as much as greenhouses with only 10 percent of the water, but many organizations are pushing the envelope in growing technology. For example, RotterZwam, which specializes in mushroom production, is growing pink oyster mushrooms in a renovated swimming pool in Rotterdam, and the PlantLab in Den Bosch is growing corn in a three-story-deep basement.

Flower gardens surround government buildings in The Hague.

The government is on board with furthering urban agriculture. The Netherlands' agriculture programs emphasize education, and the Dutch minister of

agriculture approaches the movement realistically, acknowledging that urban farming will not feed the entire country but can symbolize a connection with the earth—large-scale, high-tech solutions might not be sustainable, but they start a dialogue between producers and consumers.

The Dutch have also done a lot to combat food insecurity. The cities don't have food deserts, unemployment is relatively low, and they export many crops. However, some farmers are diversifying business to provide social care, such as teaching mentally or physically challenged people how to work in agriculture.

BOSNIA AND HERZEGOVINA

To the extent that the Netherlands' urban gardens are a way to play with the latest technology, the war-torn nation of Bosnia and Herzegovina gardens for survival. Despite facing a 57 percent unemployment rate, the Bosnian government has no plan for urban agriculture in Sarajevo; however, a few dedicated nongovernmental organizations (NGOs), such as American Friends Service Committee (AFSC) and Seeds of Peace, support multiethnic urban gardening for war refugees, veterans, and survivors. Onions, carrots, and peppers bring Serbs, Croats, and Bosniaks—Catholics, Christians, and Muslims, respectively—together in safe therapy gardens, helping them to find solace by building new skills and feeding their families.

A cabbage garden in an urban space in Livno, Bosnia and Herzegovina.

Many returnees, refugees who are returning after years away from their homes, suffer from post-traumatic stress disorder (PTSD) and other physical and mental health conditions. AFSC and Seeds for Peace initiated projects via the Community Gardens Association to provide seeds and tools to get gardens started in Sarajevo to help these people. AFSC director Davorin Brdanovic states that the mission of the gardening program is to build trust among Bosnia's ethnic groups; to provide materials, work, and therapy to low-income families; and to educate its participants about organic farming and environmental protection. More than 2,000 individuals are feeding their families and regenerating their self-worth through gardening, thanks to these programs, but thousands of hopeful participants remain on waiting lists.

Native flowers grow in Accra's arid climate.

GHANA

Urban agriculture in Accra, Ghana, is a great example of a community putting permaculture principles to work. Market gardens in the city, located on marginal land and irrigated with filtered wastewater, supply up to 90 percent of the city's fresh vegetables while at the same time drastically improving the city's ecological and socioeconomic conditions.

Because the land used for these gardens is often along gullies, drains, and bodies of water, it's not suitable for development. Placing small gardens on these sites minimizes water transportation and stabilizes the land. By collecting rainwater and using simple filter systems to clean gray water for irrigation, gardeners are improving the city's faulty wastewater management and providing much-needed ecosystem services. They're also composting for fertilizer and working with industry professionals to make fencing and farming tools from recycled organic material.

The socioeconomic edge is an advantage as well. Because lower-class citizens are selling to the middle class, there's more interaction between them, blurring class lines. Market gardens are flourishing throughout Accra, thanks to organized networks of women's groups and producers, such as the Association of Market Gardeners, which have powerful influence over local laws. As a result, the gardeners enjoy unquestioned use of vacant land and exemption from fines and taxes.

INTEGRATING WITH COMMUNITY

Fitting in, whether you move to a new area or just try something new with your front yard, is a point of discomfort for many new farmers. It's only human that we want to feel accepted and welcomed. Sometimes this is a challenge as we bravely step into atmospheres ingrained with someone else's way of doing things.

We can attempt to bring our beliefs and values into these situations while respecting traditions and acknowledging human needs. We all feel like outsiders at some point in our lives, and remembering that can help us relate to others. To repeat the quote from Empowering Women Veterans that Alvina Maynard shared: "We cannot influence change by damning our neighbors."

ISOLATION

We cocreate our communities, whether through a very intentional design process or through small adjustments to a system. Intentional communities offer wonderful possibilities while requiring intensive attention to the relationships. I've lived in a few myself, and I know it is possible for them to function well.

Melissa says, "Nobody really brought cookies to me when we moved [in]. For a while, we were the 'hippies on the hill.' We didn't drive, and we biked everywhere, so it was a sight on the road." Now, several years later, she feels very at home in her space, regularly connects with most neighbors, and feels that the folks in her area are a great resource.

Support from family, friends, and neighbors is a valuable resource.

Aunt Judy advises, "One of the most important things you need to do is get a support system. If you don't have a support system, you don't have anyplace to go when you need it." But it doesn't happen without some effort. Judy says, "It's not something that happens; it's something you do. I did not realize that this was my obligation. I just expected it to come to me. We're not meant to do it alone. But, in America, we think we can do everything on our own. We're wrong, and it's been proven over and over again."

Judy bases her outlook on her faith in God and the strength she finds in her spiritual beliefs. She says, "God didn't make one person; He made two. We can do a lot, but we still need support. When you succeed, what have you really done? You're up there by yourself. If we're growing as a group, we're succeeding together. When I succeed, we succeed, and that's where fulfillment comes in. That's what I wish someone had told me: find a support group, a church, somewhere where you can be real and honest [and] interact with someone on a real level."

A NEW NEIGHBORHOOD

When Rachael Dupree moved, she felt confused about shifting from city to country life. As a shy person, she said, "It's easy to come here and not talk to anybody." She felt as if she stuck out like a sore thumb and was uncertain about everything. She worried, "I have all these questions. How are they going to view me? How can I do what I want to do on my land and be accepted if it's different from what they are doing?" Then she caught herself with these preconceived notions of what her new community's preconceptions might be, and she tried to laugh it off.

Nevertheless, neighbors are sometimes the hardest people to get along with, regardless of the urban or rural aspect. Rural communities can be more welcoming than urban; in places full of people, it is easy to feel alone and disconnected. From my experience of living in eight different states—in cities, in suburbs, and in the country—people everywhere are just people. It's always a mixed bag, and there's always something to like about them.

Every farm has its own culture. Lisa Munniksma gently warns that community building happens naturally and in a microcosm with other farmers. She says, "When you spend this much time with people, doing things with your hands, you get to know yourself, but you really get to know other people. This is more than 'Oh, what's your favorite band? What's your favorite color?' These things don't matter—at all," she laughs. "You really get into the heart of people's histories and where they're headed."

In an urban-farm environment such as Remy's, your farm becomes part of the neighborhood.

Running away from dealing with people by going to work on a farm is ultimately counterproductive. Willingness to work through difficulties with others is a valuable life lesson. Lisa says, "Being able to accept who people are and where they are is a really vital part of being able to work with people in general and a large part on farms in particular."

A few months after Rachael shared her concerns, I visited her

Community exists even in rural areas where neighbors live at a distance from each other.

farm, and she came driving up on an ATV with a load of canned goods her neighbors had gifted to her. She had gotten to know the couple on the next farm, elders who were happy that somebody with spunk and energy had bought the land next to them and eager to help out in any way they can. Later that day, another neighbor saw us by the road and stopped to introduce herself and extend an invitation to drop by her place sometime. Rachael is beginning to feel welcome and accepted here.

Helen Terry wouldn't trade her country friends for anything. She says, "I'm fine with being out in the country. The friendships I'm developing seem to be very fast, true friends, compared to living in Houston for twenty years. I made a lot of acquaintances, and maybe had fun going out for coffee or a meal with them, but for me, friendships that I'm making out here are much more authentic, genuine, and true."

She compares the investment of time and energy she put into her connections in the city with those in the country and is grateful that the country relationships have surprised her with their generosity. She says of her city friendships, "How many of those friendships really have fruit that can be harvested when the chips are down? To me, friendship [means that] when you really need help, someone will be there."

Helen also wants to integrate her holistic fitness programs into the farming community. She sees the future of Soma Ranch as a bridge to benefit her local community in a very direct and specific way. She is exploring yoga targeted for men, but really for anyone who does heavy labor or puts in long hours of work that can strain their back, legs, and shoulders. She shares her vision: "I have this dream someday of having Soma Ranch full of local farmers, maybe even using country and gospel music and doing yoga postures that help them with their flexibility and strength but also giving them a social connection at the same time."

ORGANIZATION

Community consciousness of the value of your work is essential to continuing without burnout—whether you follow in agribusiness footsteps or carve your own new path, whether you follow an ancient path or weave many and seemingly divergent paths into a new creation that is the most viable, sustainable, healthy, and restorative for you.

Join associations that resonate with you. Many agribusiness groups have added women's versions of their good ol' boys' clubs. If following in those footsteps is valuable to you, helps you find your worth as a female farmer, and validates your work, go for it.

Do you care about getting trophies and medals for the animals you raise? If so, make yourself comfortable in the conventional ag arena. The best thing you can do for the field is bring a cooperative, feminine point of view to a competitive lifestyle. That doesn't mean you are soft, subdued, and demure; it means you have a unique opportunity to open minds and hearts and explore different approaches to animal husbandry, humane treatment, and ecological links that affect the food we eat.

The provision of separate women's groups at first riles me up, and I react with "Why can't we just go to the men's tractor information field day?" Then I think back to a women's backpacking workshop I attended when I was recently divorced and hated

The Southeast Wise Women group offers educational events and programs with its guiding principle—"Honoring Women and the Earth"—at the fore.

the idea that being partnerless would keep me from enjoying the great outdoors. It was such a relief to be around others who were there because they just needed a little boost at feeling empowered enough to do what they loved without worrying about keeping up the pace or looking a little bit amateur.

Masculine, yang structures are in place throughout agriculture, and we are so used to them in our culture that it's hard to even see them. Jessica Ballard recalls being in the first class of students to go all the way through the sustainable agriculture program at her university, and she got a lot of pushback about what sustainable farming meant to existing farmers. The reaction was, "What? We're not sustainable? We've been farming all our lives!"

It took some time, but Jessica learned to receive their reactions with empathy. She says, "I had to get over myself. When I first came into college, I was super-idealistic. I only hung out with people who agreed with me." She began to realize that there was a reason that farmers weren't taking her seriously—she had only a few years of experience. Jessica admits, "I had to really step back and be more compassionate, and hopefully folks [would] be more compassionate with me. We're not going to improve things if everybody just wants to be right all the time and nobody wants to talk and understand each other."

This is an example of allowing a feminine influence into the dominant paradigm. Women in organic and sustainable agriculture are making waves, not violent storms, by looking at a more holistic approach. As Jessica puts it, "We want everyone at the table thinking about all of the pieces instead of just the product."

We don't have to look far for support these days. Most states have an organic association, extension agencies, land grant colleges, and regional initiatives to help small farmers. When you exhaust the local resources, look to ATTRA, NRCS, Sustainable Agriculture Research and Education (SARE), and the additional resources these organizations will lead to. Facebook and other social media are ever-changing resources to connect with your local farmer friends as well.

NATURAL FRIENDSHIPS

Melissa Calhoun reminds us that the community we belong to is more than human. Far from her suburban Atlanta upbringing, she feels peaceful surrounded by the wildflowers and trees of Appalachia. Since she's been on her property so long, she says, "It feels like a friend. We know each other." When she is gone for a while, she says, "I realize some part of the place has been missing me because I've been away from it. I think it's better when we're in contact. It's good to walk around and say hi. It's good for me."

MASCULINE AND FEMININE APPROACHES

Aunt Judy explains what it was like to work with a dominant culture that did not respect her at first. "[Eventually], they saw that I was working with them, and I wasn't just up there lording over them. I wasn't just telling them what to do; we were working together." She says it opened her mind as well. Speaking about one farmer in particular, she says, "I learned a lot from him because he was really good with animals."

That wisdom of understanding a natural instinct helps Judy navigate difficult situations skillfully. "When I go into an environment that's been run predominantly by males, do I just bust in there and tell them what's what? Or do I watch and see who's

Marlen (left), Lizzy (middle), and Billie Jean (right) value their connections with other people as well as with nature.

readily available to hear? At least to be open? I go to them, and I begin to develop rapport. We all have to develop trust with each other. There's going to be some that will never trust you, and that's OK…I can't just go in there and ramrod them. You don't do that [on] your farm. If you do, you'll never have trust, even with your animals."

The reality of working in some conditions is that no matter how harmless your intentions are, and how sincerely you wish to work with them, sometimes you just need to have a good guy on your side. Judy was fortunate that her husband, James, bridged the gap. There were men who would show her respect, but only on the surface, and their pride never let them get over the fact that a woman was telling them what to do.

She remembers a situation that brought all of the tension to a head. There was a big bull that they needed to move. Judy's idea was to run a couple of cows, and the bull would follow. She discussed it with James, and he agreed. Judy says, "I went down to tell the guys, and they said, 'No, we rope the bull and drag the bull.' I said, 'Why do we rope the bull and drag the bull when we can just walk him in there with some ladies?' 'No, no, no! We always rope the bull and drag.' There was a big standoff, and I thought, *What recourse do I have?* They usurped my authority, and the tension was high."

Eventually, she called James. He came out and told them that she was the boss because he had made her the boss. Some didn't like it, but he established it fairly. Judy knows that this is not an ideal vision of women making progress in equality, but it gained her some ground where she needed it. She says, "I had an advocate, and I think you need to have an advocate until you develop the rapport that you need."

BALANCE AND THRIVE

I began this book with a look at why this is a pivotal time in herstory to become a female farmer. As I've explored this from the perspectives of women who are doing this work today, there are good reasons to think that women in agriculture will not be seen as anything unusual in the coming decades. As older generations of farmers die out, along with them go the ideals of conquering and subduing the land to force it to produce. At the same time, young women are included in all sorts of science, technology, and engineering programs alongside young men, and their fathers and mothers leave the doors wide open so they can explore the natural world and create their own tools. I see a future in which a book like this one will be unnecessary. "Woman" in *The Woman Hobby Farmer* could be replaced with "Human," for those who express humanity in all aspects of tending to the cultivation of food.

CONVENTIONAL COMMON GROUND: LEARNING FROM CONVENTIONAL FARMERS

Sustainable. Locally grown. Organic. Pesticide-free. Heirloom. Biodynamic. Free-range. Non-GMO. As a new generation of conscious farmers, we are proudly wrapping our products and practices in the labels that mean earth-friendly, healthy food. Little by little, government agencies are supporting, even embracing, the movement toward growing quality over quantity.

With the bit of power that sustainable growers now enjoy comes a warning: don't let it go to our heads. Patting ourselves on the back for all the good we are doing can polarize us greenhorns from conventional farmers. If we keep on thinking we're fighting the good fight, we'll continue to see other farmers as enemies.

Of course, we stand apart for a reason. We don't want to repeat mistakes of the past or unethically apply technology for technology's sake. But let's be careful not to overlook potential allies.

Just for the sake of exploring this topic from a new perspective, let's begin by assuming the best in others. Start from the basis that we all want to survive on this planet, and we want to cooperate with each other. Here are three snapshots in agricultural time to help us find our common ground.

Fast-forward to the future. We all want to leave a healthy world for our children. We are all trying to feed a growing world while consuming fewer resources. Faced with shocking statistics on population explosions, agriscientists are designing all kinds of innovative ways to keep us all nourished. Likewise, because we want to take responsibility for feeding our communities, urban homesteading and garden farms could take over suburbia in the near future. Yet the need to feed a growing population is nothing new.

Rewind 12,000 years. Farmers have been cultivating crops on this continent since 10,000 BC, when the shift from migratory lifestyles to more permanent settlements allowed populations to increase.

Selective breeding—taking seeds from the best crops to make even better crops—began that long ago as well.

Take a look at the present moment in history. Conventional farming has accelerated the natural progression of human technology. Sustainable farmers use ancient technology to save seeds of heirloom varieties so that fewer inputs are needed. We focus on the human–nature relationship and ensure that soil health takes top priority. This is where the divide begins.

Consider permaculture, or "permanent (agri) culture," as a way to mend the divide. A foundation of permaculture is redundancy—many elements serving the same function, overlapping roles so that failure of one part will not mean a complete breakdown of the entire system.

Organic, sustainable agriculture also uses a diverse system of plants, animals, microorganisms, water sources, heat sources, checks and balances, energy flows, moon cycles, and limitless other elements to ensure that the functions of the system are met and maintained. Like conventional farmers, this is one thing that permaculturists do well: cover all their bases.

In general, farmers do not choose their profession because they want to limit themselves. Whether conventional or sustainable, the job description is multifaceted and requires creative and critical thinking. Troubleshooting, repurposing, observing, and making adjustments—whether by tuning in to natural cycles or to the latest technology—farmers of all kinds can find common ground in their innovative and open-minded adaptations. Let's take this attitude forward as we cultivate nourishment in our unique way and work toward sharing solutions with our fellow farmers.

REFERENCES AND RECOMMENDED READING

The AppleSeed Permaculture Blog; "Eight Forms of Capital: A Whole System of Economic Understanding," blog entry by Ethan Roland and Gregory Landua, December 17, 2016.

Bane, Peter. 2012. *The Permaculture Handbook*. Gabriola Island, British Columbia: New Society Publishers.

Booth, Rebecca. 2008. *The Venus Week*. Philadelphia: Da Capo.

Bradbury, Zoe Ida, Severine von Tscharner Fleming, and Paula Manalo, eds. 2012. *Greenhorns: The Next Generation of American Farmers*. North Adams, MA: Storey Publishing.

Browning, Frank. 2016. *The Fate of Gender: Nature, Nurture, and the Human Future*. New York: Bloomsbury.

Byczynski, Lynn. 2008. *The Flower Farmer: An Organic Grower's Guide to Raising and Selling Cut Flowers*. White River Junction, VT: Chelsea Green.

———. 2013. *Market Farming Success: The Business of Growing and Selling Local Food*. White River Junction, VT: Chelsea Green.

Costa, Temra. 2011. *Farmer Jane: Women Changing the Way We Eat*. Layton, UT: Gibbs M. Smith, Inc.

Dawling, Pam. 2013. *Sustainable Market Farming: Intensive Vegetable Production on a Few Acres*. Gabriola Island, British Columbia: New Society Publishers.

Ellis, Barbara. 2012. *Starting Seeds: How to Grow Healthy, Productive Vegetables, Herbs, and Flowers from Seed*. North Adams, MA: Storey Publishing.

Ferrell, Ann K. 2013. *Burley: Kentucky Tobacco in a New Century*. Lexington, KY: University Press of Kentucky.

Gough, Robert and Cheryl Moore-Gough. 2011. *The Complete Guide to Saving Seeds: 322 Vegetables, Herbs, Flowers, Fruits, Trees, and Shrubs*. North Adams, MA: Storey Publishing.

Grandin, Temple. 2006. *Thinking in Pictures: My Life with Autism*. New York: Vintage Books.

Grandin, Temple, with Mark Deesing. 2008. *Humane Livestock Handling: Understanding Livestock Behavior and Building Facilities for Healthier Animals*. North Adams, MA: Storey Publishing.

Heistinger, Andrea (in association with Arche Noah and Pro Specie Rara) and Ian Miller (trans.). 2013. *The Manual of Seed Saving: Harvesting, Storing, and Sowing Techniques for Vegetables, Herbs, and Fruits*. Portland, OR: Timber Press.

Hurst, Janet. 2014. *The Farm to Market Handbook: How to Create a Profitable Business from Your Small Farm*. Minneapolis: Voyageur.

Kaplan, Rachel, and K. Ruby Blume. 2011. *Urban Homesteading: Heirloom Skills for Sustainable Living*. New York: Skyhorse Publishing.

Kivirist, Lisa. 2016. *Soil Sisters: A Toolkit for Women Farmers*. Gabriola Island, British Columbia: New Society Publishers.

Kivirist, Lisa, and John D. Ivanko. 2015. *Homemade for Sale: How to Set Up and Market a Food Business from Your Kitchen*. Gabriola Island, British Columbia: New Society Publishers.

Levatino, Audrey. 2015. *Woman-Powered Farm: Manual for a Self-Sufficient Lifestyle from Homestead to Field*. Woodstock, VT: Countryman Press.

Levatino, Michael, and Audrey Levatino. 2011. *The Joy of Hobby Farming: Grow Food, Raise Animals, and Enjoy a Sustainable Life*. New York: Skyhorse Publishing.

Mitchell, John Hanson. 2014. *A Field Guide to Your Own Back Yard*. Woodstock, VT: Countryman Press.

Murphy, Antonia. 2015. *Dirty Chick: Adventures of an Unlikely Farmer*. New York: Gotham.

Poppin, Jeff. Biodynamic Farming Workshop. Foxhollow Farm, KY, February 2013.

Rankin, Lissa. 2010. *What's Up Down There?: Questions You'd Only Ask Your Gynecologist If She Was Your Best Friend*. New York: St. Martin's Griffin.

Rosas, Debbie, and Carlos Rosas. 2004. *The Nia Technique*. New York: Broadway Books.

Ryle, Robyn. 2015. *Questioning Gender: A Sociological Exploration*. 2nd ed. Los Angeles: Sage.

Sachs, Carolyn E. 1983. *The Invisible Farmers: Women in Agricultural Production*. Totowa, NJ: Rowman and Allanheld.

Sachs, Carolyn E. et al. 2016. *The Rise of Women Farmers and Sustainable Agriculture*. Iowa City: University of Iowa Press.

Shiva,Vandana. "The Future of Food." Lecture presented at University of Kentucky, Lexington, KY, February 2013.

Simos, Miriam (Starhawk). 2004. *The Earth Path: Grounding Your Spirit in the Rhythms of Nature*. San Francisco: Harper Collins.

Southeast Wise Women's Herbal Conference. Black Mountain, NC, October 2016.

Stanny, Barbara. 2014. *Sacred Success: A Course in Financial Miracles*. Dallas: BenBella Books.

Steward, Keith. 2013. *Storey's Guide to Growing Organic Vegetables and Herbs for Market*. North Adams, MA: Storey Publishing.

Tomolonis, Andy. 2014. *Organic Hobby Farming: A Practical Guide to Earth-Friendly Farming in Any Space*. Irvine, CA: Lumina Media.

Wilson, Gilbert. 2005. *Native American Gardening: Buffalobird-Woman's Guide to Traditional Methods*. Mineola, NY: Dover Publications.

Woginrich, Jenna. 2013. *One-Woman Farm: My Life Shared with Sheep, Pigs, Chickens, Goats and a Fine Fiddle*. North Adams, MA: Storey Publishing.

INDEX

A

agrarian communities, 204
agricultural cooperative extension office, 7, 14, 76, 169, 182, 186–187
agricultural experiment stations, 186
Agriculture Department (USDA). *See* US Department of Agriculture (USDA)
agritourism, 145, 215
agroecologists, 89
airbnb.com, 32
Alibates Flint Quarries National Monument (Texas), 60
allergies, 82
Alternative Farming Systems Information Center (AFSIC), 14
alternatives to land ownership
  building community first, 45
  community gardening, 43–44
  neighbor agreements, 44
  personal examples, 45–47
  renting, 43
  working for land, 44–45
American Friends Service Committee (AFSC), 219
American Gut Project, 83
AmeriCorps, 59
Animal, Vegetable, Miracle (Kingsolver), 30
animals
  communicating with, 90–91
  diverse reasons for keeping, 91–92, 95
  farming and, 81, 86–87, 92
  flow of, 36
  handling animals, 201–202
  inventory of, 41–42, 87
  rescued animal, 46, 92
Animals in Translation (Grandin), 90
annual reassessment, 13
annuals (plants), 78, 98
apprenticeship programs, 18, 44–45, 174, 176–177
Arbor Day Foundation, 105
Argentina, 217
arthritis, 129
assessments
  farming for one, 13–17
  finances, 52–62
  knowledge (*see* knowledge assessment)
  mission statement, 15
  overview, 10–11
  permaculture, 28–29
  physical considerations, 47–52
  planning process, 11
  reassessment of focus, 13
  relationships and farming, 17–21
  support organizations, 14
  values, expression of, 22–23
  what you have, 34–35
  your wants (*see* wants assessment)
  . *See also specific assessment exercises*
assets, 63–67
Association of Market Gardeners, 220
associations, joining, 224–225
ATTRA organization, 62

B

balanced farm, ecological concerns, 87–95
Ballard, Jessica
  feminine farming perspectives, 225
  on feminine vibe, 80
  GreenHouse17, 32, 67, 135, 198–199
  healing practices, 134
  on idealism, 10
  personal farming story, 15, 27
  training, formal and informal, 181, 182
Bane, Peter, 40, 57, 107
barefootfarmer blog, 76
Barr, Adam, 204
bartering system, 59–62
Berry, Wendell, 33
biodiversity, 74
The Bio-Integrated Farm (Jadrnicek), 31
biological properties, of soil, 76
bluegrass, 111
Bluegrass Domestic Violence Program, 32
body awareness guide, 49–50
Bolin, Marlena, 16
Booth, Rebecca, 138
Bosnia, 219
The Bountiful Solar Greenhouse (Smith), 98
Brdanovic, Davorin, 219
breathing, self-care and, 130–133
Buffalo Bird Woman (Hidatsa tribe), 145
building community, 45, 61, 203–204
Burley: Kentucky Tobacco in a New Century (Ferrell), 100
Burnett, Graham, 58, 59
buying versus renting property, 43

C

Calhoun, Melissa
  apprenticeship programs, 18
  career description, 32, 212
  community building, 199–200
  decision making, 207
  education, 32–33, 180, 181
  farm tools, 101–103
  land assessment, 47
  natural friendships, 226
  walk-behind tractors, 106
Canadian Healthy Infant Longitudinal Development (CHILD) Study, 82
capital, types of, 59, 63–64
career choices, 32–33, 212–213
Carroll, Delia, 60
Centers for Disease Control and Prevention (CDC), 83
chemical properties, of soil, 76
"chop and drop" (cover crops), 169
Ciccarelli, Margrit, 215
climate battery, 119
Clow, Sara, 113
Coleman, Eliot, 181
Common Treasury Farm (Oregon), 46–47, 48, 174, 176

community
  building, 45, 61, 203–204, 222
  ecosystem, 142
  equestrian, 204
  gardens and gardening, 43–44, 183–184
  importance of, 5–6
  integration, 220–225
  support, 156
Community Farm Alliance (Kentucky), 16
Community Gardens Association, 219
community-supported agriculture (CSA), 16, 61, 155–156
conferences
  blog post, 30, 112–113
  "Intensive Vegetable Production on a Small Scale," 31
  knowledge assessment, 26–27, 30–31
  online forums, 32
  "Permaculture Designs for Small Farms," 31
  preparation and expectations, 26–27
conscious awareness, 131–132
Conservation District (grant program), 139
conventional common ground, 228–229
Cook, Frank, 180
coppicing, 105
couchsurfing. com, 32
cover crops, 169
Covey, Stephen, 15, 86
cross of Lorraine, 148
crowdsourcing, 83
cultivation
  animals, 81, 86–87, 92
  daily maintenance, 79
  no-till farming, 79–80
  plants, 84–85
  soil (*see* soil)
cultural capital, 63
cutting tools, 103, 105
cyclicality, 134, 138–139, 196, 199–200

D

daily maintenance, 79
damselflies, 148–149
Dancing Rabbit Ecovillage (Missouri), 45
data collection, 40
Dawling, Pam, 31
decision making, 206–207
decomposition, 207–208
Defense Department, US, 105
discouragement, unconscious, 204
diseases
  biodiversity and, 74, 82–83
  healing traditions and, 124
domestic violence, 198
dragonflies, 148–149
Ducks Unlimited, 115
Dupree, Rachael, 17–18, 72, 200–201, 207, 222–223

E

Earth Tools, 101–102
EcoFarm Conference (California), 26
ecological concerns, 88–89
ecosystem community, 142
Ecosystems Design, Inc., 119
education. *See* knowledge, learning and; lessons learned
Elliott, Kaitlyn, 67, 188–190
Empowering Women Veterans, 220
Environmental Protection Agency (EPA), 164
Epidemic of Absence: A New Way of Understanding Allergies and Autoimmune Diseases (Velasquez-Manoff), 82
equestrian community, 204
"European Year against Food Waste," 164
experiential capital, 63

F

4-H clubs, 191
4th Street Farm, 166, 170
fair share, 28
Faith Feeds (now Glean KY), 165
Farm Hack website, 14
Farm Service Agency (FSA), 14, 114
farm work guide, 51
farmers' markets, 165
farming for one, 13–17
farmland restoration, 114–117
farms and farming
  animals, 81, 86–87, 92
  basic design, 70–72
  diversity of, 6–7
  ecological concerns, 87–95
  feminism and farming, 80, 212, 225, 226–227
  gender difference in, 79–80, 93, 125, 142
  human-friendly spaces, 142–147
  livestock farming, 23, 202–203
  natural fiber farming, 22, 92, 95, 174, 184
  no-till farming, 79–80
  organic farming, 77, 183
  plants, 84–85
  priorities, 70–72
  relationships and farming, 17–21, 209
  site-specifics, 72–73
  soil (*see* soil)
  suburban farming, 215
  tools, 95
  urban farming, 213–220
  volunteering and, 136–137
  . *See also* life, integrating farming with
farmstayus.com, 27
fashion
  footwear, 96–97
  gloves, 97, 100
  slow fashion movement, 97
feminism and farming, 80, 212, 225, 226–227
fermented foods, 83
Ferrell, Ann K., 100
fiber and textiles, 22, 92, 95, 174, 184
field trips, 113
filberts (hazelnuts), 104
finances
  assets, 63–67
  bartering, 59–62
  capital, 59, 63
  cash flow tracking, 58

coaching services, 56–58
handling money, 56
monthly financial flow, 54
net worth worksheet, 55
overview, 52–53, 58
resources, 53–56
tracking income and expenses, 58
wealth, 58–59
Fisher, Rick, 118
floating wetlands, 88
flows
of animals, 36
cyclicality and, 138–139
financial, 54, 58
land assessment mapping, 36–37
foam-rolling massage, 129
Food and Agricultural Organization (FAO), 85
The Food and Heat Producing Solar Greenhouse (Yanda & Fisher), 118
Food and Justice conference (New Mexico), 160
food
desert, 157
hubs, 158
pantries, 157–158
footwear, 96–97
The Forest Garden Greenhouse (Osentowski), 99
forest gardening, 98
Forest Retreats (Missouri), 45
forests, 74
freelance farming and writing, 212
Freeman, Laura, 44
fusion-fitness program, 129
Future Farmers of America (FFA), 191
future outlook, 188–191, 227, 228

G

Gahn, Carolyn and Jacob, 114
Gaia's Garden (Hemenway), 107
Geers, Billie Jean, 96
gender differences
community building, 203–204
decision making, 189
educational opportunities, 191
in farming, 79–80, 93, 125, 142
future outlook, 227
leadership approaches, 178
male/female approaches, 226–227
male/female coworkers, 204
overcoming stereotypes, 190
social movements and, 198
women in subjugated positions, 197–199
Geological Survey, US (USGS), 73, 115
getting started, 118
Ghana, 220
"Glass Houses" (Osentowski), 107
Glean KY (formerly Faith Feeds), 157–158, 165
gloves, 97, 100
Godino, Jessica, 138
Google Sheets, 58
Graber, Jacob, 45
Graham, Jerred, 44–45
Gram Vikas Navyak Mandal Laporiya (organization), 89
Grandin, Temple, 90
Greenhorns website, 14
GreenHouse17 organization, 32, 67, 135, 198
greenhouses, 98–99, 107, 118–119
Grow Appalachia, 135
Grow Food Carolina, 113
Guthrie, Lizzy, 189, 214

H

Hammond, Marlen, 189, 204, 215
hand tools, 102
The Handbook of Permaculture (Bane), 40
harvest
assessment exercises, 157, 158, 161
community supported agriculture, 155–156
hands-off gardening, 166–171
overview, 154
parties, 155
seed sharing, 159–162
ugly food, 164–165
waste, 156–158, 164–165
have assessment, 34–35
hazelnuts (filberts), 104–105
"Head, Shoulders, Knees, and Toes" (song), 127
healing traditions, 122
Healing Wise (Weed), 122
heat, maintaining in a greenhouse, 119
heirloom seeds, 159, 161, 171
Heistinger, Andrea, 171
Help Exchange, 62
Hemenway, Toby, 107
Hendrych, Remy
annuals and perennials, 78
breathing, 130–132
cyclicality, 134, 196
on decomposition, 207–208
farming challenges, 178
fermented food, 83
sacred spaces, 142, 147
as teacher and student, 180
urban farming, 215–216
Urban Ninja Project, 7, 131
Hensley, Tim, 45
herbal medicine, 150–151, 180
heroic healing tradition, 122
Herzegovina, 219
hierarchy of food recovery and waste redirection goals, 164
Hipp, Janie Sims, 112
Hitt, Alex, 112
Hobby Farms (magazine), 107
Hoffman, Robert, 114, 116
holistic approaches, to self-care, 129–133
hormones, 138–139
hugelkultur (raised bed), 215
human-friendly spaces in farms, 142–147

I

idealism, 10
illnesses, biodiversity and, 74, 82–83
incubator schools, 113
infrastructure challenges, 16–17
Instituto Terra, 88
intellectual capital, 63
"Intensive Vegetable Production on a Small Scale" (conference presentation), 31
Intermarché (French grocer), 165

inventory, taking, 39–42
isolation, 221

J

Jadrnicek, Shawn, 31

K

Kaplan, Rachel, 60
Kavasch, E. Barrie, 147
Kentucky Artisan Barter (Facebook), 61
Kentucky FarmStart Course, 187
Kentucky Women in Agriculture, 139
Kingsolver, Barbara, 30
Knight, Rob, 83
knowledge
- apprenticeship programs, 18, 44–45, 174, 176–177
- assessment, 23–27, 30–33
- assessment exercises, 179, 185
- community gardens, 183–184
- cooperative extension service, 186–187
- future outlook, 188–191
- leadership and mentorship, 177–179
- observation skills, 37, 39, 180
- overview, 174
- students, 180–185
- . *See also* lessons learned

L

Lancaster, Brad, 89
land
- alternatives to land ownership, 42
- flows, mapping, 36–37
- inventory, taking, 39–42, 116
- mapping overview, 35–36
- prices, 43
- property layers, mapping, 38
- renting versus buying, 43
- site-specific focus, 72–73
- soil (*see* soil)
- tending with neighbors, 43
- watershed, 116
- zone mapping, 70–72

land grant institutions, 186
landscaping, natural, 108–111
Landua, Gregory, 63
Laura's Lean Beef, 44
Lawton, Geoff, 171
leadership, mentoring and, 177–179
LED lighting systems, 218
Leffew, Ben, 116
legumes, 85
Lein, Susana, 17, 65, 79, 140–141
lessons learned
- assessment exercises, 195, 208
- blog post, 209
- decision making, 206–207
- decomposition and, 207–208
- outcomes, 207–208
- overview, 194–196
- personal stories, 196–202
- realizations, 202–206
- reflections, 196–202

life, integrating farming with
- Argentina, 217
- balance and thrive, 227
- Bosnia, 219
- career choices, 212–213
- community integration, 220–225
- conventional common ground, 228–229
- feminism, 212
- Ghana, 220
- Herzegovina, 219
- isolation, 221
- masculine and feminine approaches, 226–227
- natural friendships, 226
- Netherlands, 218–219
- new neighborhood, 222–224
- organization, 224–225
- rural and urban farms, 213–220
- South Korea, 217–218

Liloia, Brian "Ziggy," 45
livestock farming, 23, 202–203
living capital, 63
Loblaws (grocery store chain), 164–165
local agricultural cooperative extension office, 7, 14, 76, 169, 182, 186–187
long-handled tools, 102–103, 106
Lotte Mart (South Korea), 218

M

Maddock, Sherry, 166, 168, 169, 170, 171
maintenance, daily, 79
"mansplaining", 200
The Manual of Seed Saving, 171
Many Hands Mondays, 155
mapping land resources
- existing resources for, 41
- flows, 36–37
- guide, 38
- overview, 35–36
- property layers, 38

masculine and feminine approaches, 226–227
massage therapy, 129
Master Community Composter, 184
Master Community Gardener, 43, 184
material capital, 63
Maynard, Alvina
- animal farming, 81, 92
- asking questions, 182
- career choices, 213
- decision making, 206–207
- integrating with community, 220
- realities of raising livestock, 202–203
- teaching, 184, 188
- values, 22

McCarthy, Ann, 104
McNeil, Lizzy, 46
The Medicine Wheel Garden (Kavasch), 147
medicine wheels, 123
Melendez, Alice, 44–45
Melt method, to self-care, 129
mentors, 112
mentorships, 177–179
Meyer, Lee, 43
microbial intelligence, 76, 82–83
Microloan Program, FSA, 14
Microsoft Excel, 58
Midwest Organic and Sustainable Education Service (MOSES) Conference (Wisconsin), 26
migrant farm-labor, 197
mindfulness, 131–132
mission statement, 15

Morales, April, 45
Moss-Greer, Angelique "Sobande," 163
motivation, 67
Munniksma, Lisa
- apprenticeship programs, 176–177
- career choices, 212
- community building, 203–204, 222
- community support, 156
- education, 182
- as a female journalist, 200
- injury prevention, 126
- land ownership, 47
- livestock farming, 23
- overcoming stereotypes, 190
- work variety, 184

mushroom production in Netherlands, 218

N

National Center for Appropriate Technology (NCAT), 62
native plants, 108–111
natural fiber farming, 22, 92, 95, 174, 184
natural friendships, 226
natural landscaping, 108–111
Natural Resources Conservation Service (NRCS), 48, 74–76, 114, 182
Nature Conservancy, 115
Navdanya organization, 160
net worth worksheet, 55
Netherlands, 218–219
nettles, 110
networking, 15, 26–27, 72, 176, 181, 213
new neighborhood, 222–224
Nia Technique, 49, 127–130
Noé, Jody, 150–151
North American Biodynamic Conference (New Mexico), 26
Northern Michigan Small Farm Conference, 26
no-till farming, 79–80

O

observation skills, 37, 39, 90–91, 118
online forums, for conferences, 32
online information sources, 181
Organic Association of Kentucky, 33, 214
organic certification, 174, 176
organic farming, 77, 183
Organic Tilth, 174
organizations
- joining, 224–225
- supporting newbies, 113

Osentowski, Jerome, 99, 107, 118, 119
overabundance, 156–158
Ozinskas, Andrew Alexander, 180

P

Pediatrics (online version), 82
people care, 28
perennials (plants), 78, 98, 105
permaculture, 28–29, 113, 140–141, 229
"Permaculture Designs for Small Farms" (conferences presentation), 31
permaculture greenhouses, 98–99, 107, 118–119
The Permaculture Handbook (Bane), 107
Permaculture Magazine, 59
pesticides, 77
phenology, 41
Phoenix (permaculture greenhouse), 107, 118
physical assessment
- ability, 48
- body awareness guide, 49–50
- farm work guide, 51
- limits, knowing, 48
- overview, 47–48, 52

physical posture, 127
physical properties, of soil, 73–75
pick and prep/farm to table programs, 155
pin oaks, 110
pitch pine trees, 151
planning
- overview, 10–11
- projects (*see* assessments)
- tasks, 126
- what you have, 34–35
- what you know, 23–27, 30–33
- what you want (*see* wants assessment)

plant guilds, 98
plants
- annuals, 78, 98
- hands-off gardening and, 169–171
- native varieties, 108–111
- perennials, 78, 98, 105
- self-pollinated, 162
- soil and, 84–85
- transplant shock, 140–141

Plowshares Community, 44–45
Pollinator Partnership's website, 168
ponds, 146
Poppen, Jeff, 76, 79–80
Post, Jane, 45
posture, physical, 127
Price, Catherine, 183
priorities, zone mapping and, 70–72
property layers, mapping, 38
psychobiotics, 83

Q

Quay, Cindi, 150–151
Quillen, Eric, 44
Quillen, Paige, 44, 45–46

R

Rainwater Harvesting for Drylands and Beyond (Lancaster), 89
Ramsey, Dave, 56
Ratajczak, Katie, 93, 190, 197, 212
realizations, 202–206
reassessment, of focus, 13
reflections, 196–202
relationships and farming, 17–21, 209
renting versus buying property, 43
rescued animal, 46, 92
restoration agriculture, 78, 104–105

Restoration Agriculture: Real-World Permaculture for Farmers (Shepard), 104
restoration farmland, 114–117
restorative ecology, 88
rhizobia, 85
riparian buffers, 88
River Hill Ranch, 92
Roland, Ethan, 63
Roots Memphis Farm Academy, 113
Rosario, Argentina, 217
Rosas, Debbie and Carlos, 49
RotterZwam (Netherlands), 218
Roundstone Native Seed, 114
rule of four, 150
rural and urban farms, 213–220

S

The 7 Habits of Highly Effective People (Covey), 15, 86
sacred spaces, 147
Sacred Success (Stanny), 34, 57
Salamander Springs Farm, 17, 113, 140–141
Salatin, Joel, 30
Sanderson, Tyler, 115, 116, 117
Scientific American (magazine), 82, 83
scientific healing tradition, 122
Scott, Delia
  career choices, 33
  education, 181
  as an extension agent, 72–73
  on footwear, 96
  on gloves, 97
  organic farming, 183
  rural and urban farms, 214
  on sun protection, 100
  on tools, 101
  working in male dominated field, 93
Second Harvest (food bank network), 165
seed banks and libraries, 159
seed sharing, 159–162
Seedleaf, community gardening organization, 43, 183–184
Seeds of Peace, 219
self-care
  assessment exercises, 123, 128, 133, 143–145
  breathing and, 130–133
  cyclicality, 134, 138–139
  healing and, 134
  herbal medicine, 150–151
  holistic approaches, 129–133
  human-friendly spaces in farms, 142–147
  overview, 122–125
  proactive prevention, 126
  understanding our bodies, 125–130
  volunteering and, 136–137
self-pollinated plants, 162
Sequatchee Valley Institute (Tennessee), 33
Shaker Village, 116, 117
sharecropping, 44
Shepard, Mark, 78, 104–105
Shiva, Vendana, 160
Singh, Laxman, 89
slavery, seeds and, 163
SlideShare.net, 31
Slone, Storey, 15–16
slow solutions movement, 29, 92, 95
Smith-Lever Act (1914), 186
social capital, 63, 64
social permaculture, 60
soil
  animal farming and, 81, 86–87
  annuals or perennials, 78
  biodiversity of, 82–83
  biological properties, 76
  chemical properties, 76
  hands-off gardening and, 168–169
  organic farming, 77
  overview, 73
  physical properties, 73–75
Soil and Water Conservation District, 114–115
Soma Ranch (Texas), 46, 132, 223–224
somatics, 132
South Korea, 217–218
Southeast Wise Women, 122, 224
  Herbal Conference (North Carolina), 5, 26, 150
Southern Sustainable Agriculture Working Group (SSAWG) Conference (Kentucky), 26, 30–31, 112, 209
spaces
  overview, 142–143
  sacred spaces, 147
  sharing of, 143–146
spicebush tea, 151
spiritual capital, 63
stacking functions, 84–85
Stallins, Tony, 83
Stanny, Barbara, 34–35, 56
Strange, Joscelyn, 66–67, 96, 191
suburban farming, 215
sun
  as natures tool, 107
  protection from, 100–101
support systems, 221
Sustainable Market Farming (Dawling), 31
Sweetgrass Granola, 114

T

13 Moon CoLab, 60
Terry, Helen
  community building, 223–224
  farm education, 182–183
  reflections, 196–197
  rescued animals, 46, 91–92
  Soma Ranch, 132–133
thermal mass, 119
Thompson, Michael, 119
"three sisters" of the garden, 84–85
tilth, 73
Time Banks, 61
tools
  cutting tools, 103, 105
  footwear, 96–97
  for sun protection, 100–101
  gloves, 97, 100
  hand tools, 102
  invite nature inside, 98–100
  Leatherman multi-tool, 94
  long-handled tools, 102–103, 106
  simple tool guide, 101–102

sun, as nature's tool, 107
task-specific tools, 102
tool kits for women, 94
toolbox assessment guide, 95
walk-behind tractors, 106
Total Money Makeover (Ramsey), 56
transplant shock, 140–141
Trout Unlimited, 115
Tucker, James, 19, 20, 37, 65, 79, 227
Tucker, Judy (aunt)
being in sync with nature, 79
cultivation, 79
decomposition, 207–208
freedom, 207
handling animals, 201–202
mistakes and lessons learned, 205–206
observation of patterns, 90–91
observation skills, 37, 39, 180
outside expertise, 65–66
overabundance, 158
partnership values, 18–20
planning for the unexpected, 196
sharing husbandry knowledge, 178
strength of women farmers, 53
support systems, 221
value of local resources, 73
working in male dominated field, 226–227

U

ugly food, 164–165
unconscious discouragement, 204
underground food web, 74
United Nations, 85
United Way, 53, 56
University of Kentucky, 32, 43, 83, 139
U-pick programs, 156
Urban Bees Seoul (South Korea), 217
urban farming, 215–216
Urban Ninja Project, 7, 131
US Department of Agriculture (USDA), 14, 74–76, 105, 114, 164, 182, 186
US Department of Defense, 105
US Environmental Protection Agency (EPA), 164
US Geological Survey (USGS), 73, 115

V

values
expression of, 22–23
in partnerships, 19-21
Velasquez-Manoff, Moises, 82
The Venus Week (Booth), 138
vertical gardening, 218
"A Village That Planted Its Rain and Watershed" (Lancaster), 89
visioning, 39–40

W

wants assessment, 12-13, 15-23
Wartes-Kahl, Angela
career description, 212
Common Treasury Farm, 46–47, 48, 174, 176
organic certification training, 77
rural farming, 215
women's natural qualities, 202
working in male dominated field, 124–125
wasps, 167–168
waste
overabundance and, 156–158
ugly food and, 164–165
Waste Not, Want Not report, 164
wastewater management, 220
water resource assessment, 36
watershed, 116
Watershed Farm (now GreenHouse17), 135
Web Soil Survey website, 116
Weed, Susun, 122
Wigglesworth, Shayne, 129, 130
wild ginger, 109
wildlife, hands-off gardening and, 166–168
wildlife cohabitation, 146
Wildlife in Your Garden (Lanier), 41, 86, 146
wind assessment, 36
Winnie the Pooh, as agriculture diversity, 114
wise woman healing tradition, 122, 124
Women Food and Ag Network (WFAN), 14
Wood, Corinna, 122
Wood, Pamla, 139
work share/trade arrangements, 44-45, 61-62, 155
Worldwide Opportunities on Organic Farms (WWOOF), 17, 62

Y

Yanda, Bill, 118
"The Year of Mud," 45
YouTube, 183

Z

zones, land assessment and mapping, 70–72

PHOTO CREDITS

Front cover: Fotokostic/Shutterstock, Jenoche (bottom left), wavebreakmedia (bottom middle), Volodymyr Baleha (bottom right)

Back cover: Karen Lanier

Graphic elements/backgrounds: schab/Shutterstock, Volkova Anna/Shutterstock, mythja/Shutterstock

Author photo: Russ Turpin

Karen Lanier: 1 (top right) (bottom Right), 4, 6–8, 10, 11, 15, 16, 17, 18, 19, 20, 22, 26, 32, 33, 37, 39, 42, 46, 52, 57, 62, 65, 67, 68, 72 (right), 73, 78–80, 84, 86, 92, 96–98, 104–108, 111, 119, 122, 127, 131, 134–136, 142, 147, 162, 176–178, 180–184, 192, 196, 198–203, 205, 206, 208, 210, 213, 214, 216, 221, 222, 224–226, 229

Shutterstock: Johnny Adolphson, 223; Africa Studio, 190, 57; Nate Allred, 3 (bottom right); Subbotina Anna, 61; apiguide, 170; azem, 3 (bottom center); Barbara Bekei, 160 (top); bikeriderlondon, 1 (top middle), 43; Bluehousestudio, 63; bubutu, 172; Norman Chan, 170; Chutima Chaochaiya, 28; COLOA Studio, 169; Michael Courtney, 3 (bottom left); DiamondGT, 187; Dimanchik, 83; DnD-Production.com, 31; mircea dobre, 165; ESB Professional, 120; Feng Yu, 146; Fotokostic, 125; Steven Frame, 117; goodluz, 3 (top right), 124; gpointstudio, 1 (bottom left), 95; Anne Greenwood, 36; grublee, 159; Arina P Habich, 77; Hawk777, 168; LorraineHudgins, 167; MaxyM, 115; Michal Hykel, 148; Adriana Iacob, 219; igorsm8, 82; igorstevanovic, 1 (top left), 132; Kyselova Inna, 154; irin-k, 99; Ruslan Ivantsov, 138; Eric Isselee, 88, 189; Jaengpeng, 139; jamesteohart, 218 (top); Budimir Jevtic, 3 (top center), 66; Gregory Johnston, 81; Kekyalyaynen, 3 (top left); Thammanoon Khamchalee, 140; Khumthong, 163; Agatha Koroglu, 174; ksena2you, 60; Marina Lohrbach, 194; Dariush M, 197; Robyn Mackenzie, 58; MaxyM, 102; mholka, 49; Milosz_G, 44; Monkey Business Images, 152; Jorge Moro, 151; Maks Narodenko, 186; NattapolStudiO, 164; nine_far, 160 (bottom); claire norman, 90; PEPPERSMINT, 93, 120-121; Alexander Raths, 155; Nataly Reinch, 220; RTimages, 158; Andreas Saldavs, 130; Elena Schweitzer, 209; OLEKSANDR SHEVCHENKO, 72 (left); sima 1(bottom middle); stockcreations, 166; tomiimages, 92; Kelly vanDellen, 218 (bottom); vnlit, 149; Watch_The_World, 217

USDA Natural Resources Conservation Service, 74, 75

ABOUT THE AUTHOR

Karen Lanier is a naturalist, documentarian, teacher, artist, and gardener who explores the intersections of nature and culture. Karen holds degrees in photography, foreign language, conservation studies, and documentary studies as well as a professional environmental educator certificate. She worked as a seasonal park ranger in state and national parks across the United States before settling in Kentucky. Her AmeriCorps volunteer experience with Seedleaf, a community gardening nonprofit, helped shift her migratory perspective on life toward putting down roots. Karen is actively involved in the native plant, community gardening, and environmental education groups in her area. She is a regular contributor to *Hobby Farms* magazine and editor of the book *Wildlife in Your Garden* (Lumina Media, 2016).